THE EULER–MAYER CORRESPONDENCE (1751–1755)

The Euler–Mayer Correspondence (1751–1755)

A NEW PERSPECTIVE ON EIGHTEENTH-CENTURY ADVANCES IN THE LUNAR THEORY

by

ERIC G. FORBES
University of Edinburgh

American Elsevier Publishing Company, Inc.
New York 1971

Published in the United States by
American Elsevier Publishing Company, Inc.
52 Vanderbilt Avenue, New York, New York 10017

International Standard Book Number: 0-444-19580-7

Library of Congress Catalog Card Number: 75-173862

Printed in Great Britain

CONTENTS

PREFACE

The major justification for publishing an English translation of the thirty-one letters which constitute the known correspondence between the Swiss-born mathematician Leonhard Euler (1707–83) and the German astronomer Tobias Mayer (1723–62) rests in the contribution of its central theme—the theory of the Moon's motion—to the acceptance of the Newtonian philosophy during the eighteenth century. Not only was the lunar theory then being regarded as a test case for the validity of interpreting the physical world in terms of the geometrical axioms of Newtonian mechanics, but it was at the same time providing a stimulus to the development of mathematics by revealing the need for a greater degree of generality and rigour in the new analytical methods which Euler himself, Alexis Clairaut (1713–65) and Jean d'Alembert (1717–83) were among the first to adopt. The independent researches of these three men mark the beginning of a new and very fertile phase in the history of physical astronomy, which culminated with the publication of the first volume of Pierre Simon de Laplace's *Mécanique Céleste* (5 vols., Paris, 1799–1825). Newton's axioms of motion and his hypothesis of an inverse-square law of gravitational attraction provided a sufficient basis for the brilliant achievements of celestial mechanics embodied in this monumental work; thus astronomy became even more intimately linked with geometry, and the long-term stability of the solar system was assured without the need to introduce the transcendental hypothesis of a Divine Being—so necessary a feature of Newton's own providential explanation of the structure of the solar system.

Mayer's own rôle in this development has been summarised by the nineteenth-century historian of astronomy Robert Grant, on p. 208 of his *History of Physical Astronomy* (London, 1852), as follows:

> Mayer has earned for himself an imperishable reputation by his lunar tables. Some astronomers may have achieved results of greater brilliancy, but few have been privileged to contribute so directly to the well-being of their fellow-men.
>
> Even independently of the improvement which he effected in this instance, Mayer possesses claims to be ranked with the greatest astronomers of ancient or modern times; but as in the case of either of his illustrious contemporaries, Bradley or Lacaille, his labours are not of a nature to be generally appreciated, and therefore his reputation is less widely diffused than that of many an individual whose contributions to science, though more showy, are infinitely less substantial.

Although it is undoubtedly true that the complexity of Mayer's researches was a reason for the subsequent lack of diffusion of his results, it is equally significant that his pioneering efforts to improve the position of geodetic, navigational and astronomical instruments were quickly superseded in this period of rapid growth in the physical sciences. Moreover, a considerable portion of his work has never been published, and that which has been is either in Latin or in German; thus it has hitherto remained inaccessible to many potentially-interested students of astronomy in English-speaking countries.

In view of the fact that a great deal of insight into the nature of Mayer's labours can be gained from a study of this correspondence with Euler, and that Mayer's lunar tables (not Euler's) were directly responsible for the significant improvements that occurred in navigational practice during the latter half of the eighteenth century, it was decided to centre the Introduction upon the former rather than the latter. Additional factors influencing this decision were that: (i) Mayer is in any case the lesser known of the two correspondents; (ii) he wrote twenty of the thirty-one letters comprising this correspondence; and (iii) his initial researches in lunar theory were entirely dependent upon Euler's mechanical treatises, whereas the only clear evidence of a reciprocal influence is the latter's renewal of earlier investigations into the theory of astronomical refraction.

According to Sigmund Günther, who wrote the article on Tobias Mayer in the *Allgemeine deutsche Biographie*, xxi (1885), 109–16, the historian requires to take account of the latter's contributions to three distinct disciplines: mathematics, physics and astronomy. Such a classification of Mayer's writings tends, however, to suggest some kind of methodological distinction in the character of his treatises which simply does not exist. A more meaningful approach would be the straightforward chronological one, based upon the following four periods of Mayer's life: his youth in Esslingen (up to 1744); Augsburg (1744–46); Nuremberg (1746–51); and Göttingen (1751–62). Thanks to Paul Eberhardt, a former director of the Esslingen archives, the first period is as well documented as it can be; but a thorough search of the Augsburg and Nuremberg archives has failed to reveal any primary sources of information about Mayer's social circumstances during the years which he is known to have spent in these towns. However, the comparative scarcity of biographical material does not preclude the possibility of our being able to construct a consistent (if not accurate) interpretation of the motivation for his work and the nature of his scientific achievements from the internal evidence of his own writings.

This study has been made in the light not only of his correspondence with Euler, but of all the known documentary evidence by, or about, him—including a large quantity of his still unpublished treatises, lecture notes and correspondence which was allowed to lie practically neglected in the Göttingen University Observatory for over two hundred years. They were wisely transferred in the summer of 1965 to the local university library where the author has been able to acquire easy access to them. Although there are several aspects of Mayer's work which one might consider, the features which will be stressed here are that:

(a) the primary motivation for Mayer's astronomical researches was his devotion to the aim of improving the state of cartography by providing accurate determinations of terrestrial latitudes and longitudes; and that
(b) his success in constructing lunar tables which were at least five times as accurate as those of his famous contemporaries Clairaut, Euler and d'Alembert lay directly in his own personal skill in improving his observational technique—not in the greater profundity of his lunar theory nor in the superiority of the astronomical instruments at his disposal.

Mayer's researches, although they at first give the impression of being rather diversified in content, are characterised by a unity of purpose which makes it possible to judge the extent of Euler's influence upon them. To illustrate that influence, and to reveal the nature of Mayer's individual contributions to the development of the lunar theory, are the two prime aims of this book.

The translation of the correspondence itself has been made from the published copies of the German originals in the Russian periodical *Историко-астрономические исследования*, v (1959), 271–444 and x (1969), 285–310. References to the letters are indicated by square brackets; however, the numbering convention employed here is not the same as that used by Dr Y. K. Kopelevich in the earlier of the above-mentioned articles since it has had to be modified to take account of the subsequent discovery (by the author) of the four letters from Euler to Mayer printed in the later article. The bibliographical notes in Russian supplied by Dr Kopelevich herself, which Prof. A. J. Meadows, Department of Astronomy, University of Leicester, kindly translated for my benefit, have proved to be quite helpful, although their content has been checked independently wherever possible.

One of the best introductory surveys of developments in the lunar theory prior to the time of Newton is Hugh Godfray, *An Elementary Treatise on the Lunar Theory* (Cambridge, 1853). The historical sources on which a thorough and more comprehensive study of the lunar theory and its many associated aspects of astronomy would require to be based, are contained in J. C. Houzeau and A. Lancaster, *General Bibliography of Astronomy to the Year 1880*, edited by D. W. Dewhirst (2 vols., London, 1964). A mine of untapped manuscript material on the various astronomical methods for finding longitude at sea is presently preserved at the Royal Greenwich Observatory, Herstmonceux Castle, Hailsham, Sussex, England, and is now available on microfilm; for details, see the author's "Index of the Board of Longitude

Papers at the Royal Greenwich Observatory", *Journal for the History of Astronomy*, **1** (1970), 169–79.

It is the author's earnest hope that this booklet will encourage other historians of astronomy to explore this somewhat neglected field and thereby supply "grist to the mill" of historians of ideas who are inclined to over-simplify the eighteenth-century cultural scene by attributing too much to the direct influence of Newtonian science. As far as the question of motivation is concerned, the present study clearly reveals that Mayer's environmental circumstances, natural inclinations and urge to put German cartography on the map (as it were), had a more determinative influence upon the pattern of his astronomical researches than some innate conviction that the pathway to progress could be attained only through the application of Newtonian mechanics to the problems of the natural world.

ERIC G. FORBES

History Department,
University of Edinburgh,
Scotland
18 September 1969

INTRODUCTION

[Johann] Tobias Mayer was born in Marbach, a small town on the river Neckar only a few miles from Stuttgart—better remembered as the birthplace of the German poet, dramatist and philosopher Friedrich von Schiller (1759–1805)—on 17 February 1723. His family settled less than two years later in Esslingen, shortly after his father found employment as foreman in charge of well-digging in the service of this town. It was here that he grew up, and was destined to spend most of his later childhood and youth in the local orphanage after the death of his father in 1731. An incomplete autobiographical account[1] of his childhood prior to this time reveals the existence even in those tender years of three characteristics which are also evident in his later scientific work: a natural artistic talent, a fascination for military affairs, and an antipathy towards irrelevant or useless knowledge. It was his skill, while not yet fourteen years of age, in drawing architectural plans of the layout of the Old St Katherine's Hospital in Esslingen where his mother was obliged to work as an orderly that attracted the attention of an elderly non-commissioned officer in the Swabian District Artillery stationed in that town, and made him take an active interest in Tobias's education. By the time he was sixteen, the latter had received a thorough grounding in the principles of fortification, ballistics, etc., and had developed a liking for elementary analysis and practical geometry. Mathematics did not feature on the curriculum of the Latin school which he was now attending, but he was fortunate in receiving permission from the headmaster to read the mathematical books in his private library, and in finding in an intelligent local shoemaker someone with whom he could share his youthful enthusiasm for this subject. During the next few years, he found an outlet for his talents by giving private lessons in the different aspects of artillery, and in geometry, surveying and scale-drawing, to the sons of a number of local citizens who were destined for a military career.[2]

Mayer's first publication, completed on the occasion of his eighteenth birthday, was his *Neue und allgemeine Art, alle Aufgaben aus der Geometrie vermittelst der geometrischen linien aufzulösen* (Esslingen, 1741) which, as the title suggests, was concerned with the solution of geometrical problems by the application of analytical methods. Significantly, it culminates with a demonstration of a problem which was of practical importance in the design of military fortifications; namely, that of inscribing an irregular polygon in a circle. His *Mathematischer Atlas* (Augsburg, 1745), published a year after he had left Esslingen and taken up employment in Augsburg with the copper-engraving firm of Johann Andreas Pfeffel, is an impressive index of the extent to which he had by then developed his artistic talent and knowledge of practical geography. The sixty coloured plates which it contains present the whole range of elementary mathematics with astronomical, geographical, military, architectural and mechanical applications (in that order). The feature which merits particular attention in the present context is Mayer's familiarity with the methods of stereographic and orthographic projections, and his ability to apply these respectively to the construction of maps and to the graphical construction of solar and lunar eclipses, and to eclipses of Jupiter's satellites.

Johann Michael Franz's advertisement in 1746[3] for extra staff for the Homann Cartographic Bureau in Nuremberg to continue the tradition of the late Professor Johann Matthias Hase (d. 1742) who had re-introduced the stereographic horizontal method of map-making in maps that he had made for the Homann firm, was therefore a strong incentive for Mayer to leave Augsburg. It is not surprising therefore that when Franz gave Mayer the offer of an appointment where one of his first tasks was to assist in the final stages of preparation of the *Homannische-Hasischen Gesellschafts Atlas.*—Franz's major project at that time—he turned down Pfeffel's offer of generous terms if he were to remain in his employment and went to work in Nuremberg. This atlas, which duly appeared in 1747, contains eighteen maps of which two were constructed by Mayer himself.[4] As Franz explains in his preface to it, one reason why it had taken so long to prepare was that all observations of latitude and longitude had had to be critically examined before they were adopted, since different "authorities" quoted different results. The task had been rendered particularly difficult on account of the lack of information regarding the observer, his instruments, his observational technique, the precepts on which he based his calculations, and the scatter of the sources in which these data were to be found.

It was through his involvement in this work that Mayer came to appreciate the great need for obtaining reliable values of geographical co-ordinates, and two unpublished collections of his papers bear testimony to his efforts around this time to collate information from a wide variety of sources.[5] He was well aware, however, that this was a necessary but not sufficient condition for the successful prosecution of his task. The factor which he recognised as being lacking is revealed by the following extract from one of these, at the beginning of a section devoted to a discussion of the latitude of Nuremberg:

> It is only too well known in geography that very many, indeed most, of the numerous known and noteworthy places need to have their latitude and longitude more accurately determined. *To achieve this is one of the most important objects of astronomy*, and those who are occupied with this science have continually made great efforts to obtain the kind of observations from which the

geographical position—that is, the latitude and longitude of the place where they are—can be derived. Few, however, have extended their labours to the problem of how their own observations might properly be applied to this goal.[6]

Mayer's subsequent achievements in astronomical science can, in fact, be properly appreciated only if they are regarded as a conscious effort on his part to fulfil this urgent need, which constituted at the same time one of the primary objects of the newly-created Cosmographical Society of Nuremberg.[7]

It is in Franz's preface to the above-mentioned Homann Atlas that we find the earliest reference to Mayer's awareness that significant errors were inherent in the orthographic method of projection which he was employing in his construction of a map intended to illustrate the path of a partial lunar eclipse due to occur in the night between the 8 and 9 August 1748.[8] The basis of the calculations required for this purpose was not Mayer's own solar and lunar tables which he had employed when making similar constructions in his *Mathematischer Atlas*, but those of Leonhard Euler published in the first volume of the *Opuscula* (*varii argumentii*), **1** (Berlin, 1746) from which the solar and lunar ephemerides in the new *Berlin Calendar for 1746*[9] had been prepared. The object of the exercise was to reveal the usefulness of the application of the orthographic method of projection for the determination of geographical longitude, in accordance with one of several closely related aims of the mathematical class of the Cosmographical Society, to which Mayer belonged.[10]

In principle, the procedure for the determination of differences in terrestrial longitude from observations of a lunar eclipse was simple; one had merely to compare the local times at which the same topographical features on the Moon's surface were gradually obscured by (and emerged from) the Earth's shadow as it moved across the surface of the Moon's disc, as estimated by two independent observers situated on different meridians. The mean time-interval based on a number of such occurrences could be found and expressed in angular measure by means of the identity 24 hours $\equiv$ 360°.

There were, however, two particular difficulties which had to be overcome; one was the question of the precise value of the lunar parallax, and the other was the uncertainty in identifying the individual lunar markings. Mayer's use of an orthographic projection, with its assumption that the Moon is being viewed as if from an infinite distance, was tantamount to neglecting the lunar parallax altogether. Thus it is scarcely surprising that Mayer obtained quite different results when he compared this with an alternative construction in which the observer is supposed to be stationed at the distance of the Moon, and whose eye would have to remain in that position if he were to continue to see the Earth's apparent diameter, the sizes of the umbra and penumbra, and the apparent motions of these shadows across the Earth's surface, on the same scale. Mayer was also quick to realise that even the best of the lunar maps in existence at that time—namely, those of Johann Hevelius[11] and Giovanni Riccioli[12]—were far from reliable. For those reasons, he resolved to abandon the use of lunar eclipses as a means of finding longitude, and to prepare his own lunar map. The graphical construction of solar eclipses was also rendered uncertain by the use of an orthographic projection, but the major disadvantage was the infrequency with which they occur. Experience had shown that eclipses of Jupiter's satellites were also unsuitable for Mayer's purpose, where a high degree of accuracy was expected.[13]

The occultation of a star by the Moon, on the other hand, was a phenomenon which could be timed to within a second with the aid of a good telescope, and several occurred each month; so it appeared to offer a much more reliable means of longitude determination. If due allowance were made for the lunar parallax, and if the local apparent times of the same occultation were known from similar observations by independent observers situated on different meridians, the resulting time differences would be a direct measure of their separation in terrestrial longitude. Mayer, however, preferred to use a glass micrometer of his own design[14] to measure several small celestial arcs between the star and the Moon's limb before immersion and after emersion, in order to obtain a more accurate estimate of the times at which these two events occurred. During the years 1747 and 1748, he observed occultations of the Pleiades and a number of other stars in this manner with the aid of a 9-ft focus telescope which Franz had purchased for the Homann firm.[15]

It was as a direct result of these observations that Mayer first became convinced of the fact that the Moon does not possess an atmosphere; for he noticed that the immersions of stars occurred instantaneously and that their light rays never exhibited any diminution of intensity in the immediate neighbourhood of the Moon's limb. In "Tobias Mayers Beweis daß der Mond keinen Luftkreis habe" (1750),[16] he put forward a number of physical arguments to support this contention; the principal one being the complete absence of refraction of the starlight, which he claimed that he would have been able to detect even if there had been a lunar atmosphere as much as 120 times "more subtle" than that of the Earth. At first he was inclined to ascribe the existence of a halo round the moon—a phenomenon previously seen by several observers during the occurrence of a *total* solar eclipse—to the refraction of light through the Earth's own atmosphere, but as a result of reading Euler's "Essai d'un explication physique des couleurs engendrées sur des surfaces extrêmement minces" (1750),[17] he was later converted to the belief that this and other similar appearances might be produced by the effect of diffraction in broadening and expanding light pulses [29]. Mayer's arguments were firmly refuted in 1751 by the Berlin doctor Christlob Mylius, who had himself written a review article on this controversial subject five years previously, in which he had favoured the existence of a very rare atmosphere surrounding the moon.[18] In his counter-attack on Mayer,[19] which the latter wisely refrained from pursuing any further, Mylius quotes Euler as supporting his own point of view.[20] Thus Mayer was induced to raise this question with Euler himself [3], and was no doubt pleased by the latter's courteous reply that he merely wished to insist upon the reality of the phenomenon and not upon its cause [4]. Indeed, Euler went so far as to say that he would be willing to retract his earlier view if the results of accurate micrometric measurements of occultations (which he hoped that Mayer himself would make) should be found to require it [6].

Mayer began, early in 1748, to put into practice his intention of laying the astronomical foundations for an accurate lunar map. His time-consuming programme of micrometric observations of the lunar markings was made from the Homann buildings between April 1748 and August 1749. His technique was to align his micrometer in the focal plane of his telescope so that its parallel rulings were themselves parallel to the direction of the Moon's apparent motion; then, having mounted his telescope (with the micrometer in this position) on a machine designed by Franz which caused it to follow the Earth's rotation, he recorded the mean solar time at which the leading limb of the Moon crossed one of the perpendicular lines, noting, whenever possible, the micrometric position of the north or south limb at the same instant. The timing was done with the aid of a simple pendulum with a lead weight in a position which was convenient for him to observe, the time of whose oscillations he compared immediately before and after his observations with a seconds pendulum in the neighbouring room. The estimated times are quoted by him on some occasions to the nearest 0·1 s, which is indicative of the high precision which he was attempting to attain. When the leading limb of the Moon was not visible, he timed the passage of the horns through the perpendicular line.

From these measurements, he was able to calculate the Moon's diameter in cases where the line joining the horns was at a very acute angle to the micrometric rulings; otherwise, a little before or after making his observations, he set the micrometer so that the perpendicular line passed through the two horns of the Moon, and estimated its diameter from the distance between the parallel rulings. This enabled him now to determine uniquely the positions of the lunar markings on the Moon's disc in terms of the *relative* distance from the disc centre and the angle made at the centre with respect to the north direction. Thus the effect of the Moon's variable distance from the Earth was eliminated. The elaborate procedure by which Mayer systematically rids his results of all further dependency upon the terrestrial observer, and accounts quantitatively for the effect of the Moon's libration, is described in detail in his lengthy "Abhandlung über die Umwälzung des Monds um seine Axe / und die scheinbare Bewegung der Mondsflecken"

(1750),[21] which is noteworthy in being the first known example of the use of selenographic co-ordinates to specify the Moon's topographical features. The selenographic latitude was found by referring the positions of the lunar markings to the Moon's equator, analogous to fixing terrestrial latitudes. Mayer selected as the prime meridian of selenographic longitude, that which lies in the same plane as the line joining the centre of the Moon to the *mean* position of the Earth. Since the Earth's variable motion around the Sun was accurately known from solar tables, the relation of its mean to its true position could always be easily calculated.

Encouraged by his success in analysing the causes of libration and determining their respective effects, Mayer also embarked upon the preparation of lunar globes[22] but, for various reasons, this project never came to fruition.[23] He did, however, manage during the years 1748–50, to draw six of the fourteen segments required for a 15-inch lunar globe which were subsequently engraved on copper at the Nuremberg School of Art.[24] Although not much work would have been required to construct the globe with the aid of these and Mayer's drawings of the remaining planispheres, no one actually took up this project and it was left to the Englishman John Russell at the end of the century to achieve independently what Mayer had intended.[25]

The scholarly reputation which Mayer won for himself both through his skill and industry as a cartographer and—more importantly—through his original astronomical researches which constitute the most impressive feature of the *Kosmographische Nachrichten und Sammlungen auf das Jahr 1748* (Nuremberg, 1750), was responsible for his being called in November 1750 to a professorship at the Georg-August Academy in Göttingen. He duly took up this appointment in Easter of the following year. His title, Professor of Economy, was only a nominal one, for in his letter of appointment his duties are stated to be the teaching of practical mathematics, and research.[26] He brought with him to Göttingen his new bride, Maria Victoria Gnüge, the daughter of a former clergyman in Neckarbischofsheim and a sister-in-law of Franz, through whom they were introduced after Maria came to live with her relatives in Nuremberg following the death of her father in 1747.[27] Among the thirty or so maps which Mayer had already drawn before his departure from the Homann firm, the most important (certainly from a historical point of view) is his critical map of Germany, completed in 1750, and first published three years later.[28] This map was deliberately made, at his own suggestion, for the purpose of testing how far someone such as himself whom Franz describes as "very skilled in the art of map-making",[29] could achieve accuracy with each and every possible kind of aid available at that time. Mayer's view on this subject is evident in the following extract from his unpublished "Collectanea geographica et mathematica" (1747)—itself a product of the laborious compilation of sources required for this purpose—where he introduces the practical problem of constructing land maps:

> The geographer who shows how far he is able to extend his researches and how skilled he is in their application, has nothing further to answer for in his work than the aids which he has used and how he has assembled them. An infinite number of interrelationships can be found in the construction of a land map, which one can appreciate fully only if one follows the approach and the method of the geographer for himself.[30]

He then goes on to distinguish three types of aid for the preparation of land maps which he intended to discuss in turn; namely geometrical aids, astronomical aids, and large-scale drawings of small areas ("Spezialkarten"). The treatise is, however, incomplete and his subsequent remarks relate only to geometrical aids; in particular, to the problems of the accurate measurement of angles during trigonometric surveys, and of the estimation of distances directly measured with different and often totally unreliable standards of length.

When criticising the goniometric instruments commonly used for the former purpose on the grounds of their comparative smallness and the inaccuracy of their graduations, Mayer certainly had in mind the "astrolabe" and "recipiangle" which he had described and illustrated in Tab. XI of his *Mathematischer Atlas*. Although he was later to change the status of the astrolabe into that of a precision instrument—a forerunner of the theodolite—by screwing it on to a wooden stand, equipping it with a micrometer screw, and attaching a telescope (including a cross-wire) and a water level,[31] he appears to have initially favoured the recipiangle. In its primitive form, this latter instrument consisted of little more than two brass rulers pivoted about one end; however, by replacing one of the diopter arms by a telescope and introducing a so-called "repeating principle" (*artificium multiplicationis*) in his method of angle measurement, Mayer was able to effect a twenty-fold improvement in the accuracy of determining angular distances between two terrestrial objects (viz., from the customary $5'$ to between $10''$ and $15''$).[32] Experiments made with this new goniometer during 1753 and 1754, convinced him of the importance of two further factors in limiting the accuracy of such measurements to which he had not at first attached any importance —the defects of the human eye, and the effects of atmospheric scintillation.[33]

The problem of establishing a quantitative relationship between astronomical or terrestrial refraction and the density of the air, and his researches on the lunar theory, were the two topics discussed by Mayer when he initiated his correspondence with Euler on 4 July 1751. His own experiments had led him to believe that observations of refraction might constitute a more reliable means of determining the speed of sound than the usual method depending upon barometric experiments and Boyle's Law [1]. Euler suggested that the different results yielded by these two methods might be understood if a distinction were drawn between the propagation of a single pulse and that of a series of pulses, in accordance with the demonstration in his "De propagatione pulsuum per medium elasticum" (1750)[34] that the speed of a pulse depended only upon the density and elasticity of the air at every given point in the aetherial medium through which it passes. Euler elaborated this idea in another article entitled "Conjectura physica de propagatione soni ac luminis" (1750),[35] where he also makes the very relevant and valid point that the speed of sound may be affected by the wind velocity. This is why the value of 1040 ft/sec. which he had previously deduced in his "Nova theoria lucis et colorum" (1746)[36] could be regarded merely as a rough approximation, and why the 16% excess over Mayer's figure of 900 ft/sec. obtained from his observations of astronomical refraction was not significant [2].

After making further attempts to improve the formula for terrestrial and astronomical refraction,[37] Mayer temporarily abandoned his efforts and left Euler to develop the ideas which he had described in the *Berlin Calendar for 1751*.[38] The fundamental difficulty which the latter had been experiencing in his own fruitless attempts to derive a theoretical formula for astronomical refraction was that of integrating the differential equations so as to obtain reliable approximations for the amount of refraction at all angles of incidence of the refracted ray. Even Pierre Bouguer (1698–1758) had not, in Euler's opinion, succeeded in doing this, although his efforts had been rewarded by a prize from the Paris Academy in 1741.[39] Mayer found that he himself could obtain the best level of agreement between theory and observation by assuming that the density gradient in the Earth's atmosphere was linear, but although this hypothesis had the added advantage of permitting a simple integration to be made on his basic equation, it was "an open contradiction of the truth" [5]. Mayer's formula for refraction is appended to his letter of 6 January 1752: it was in excellent agreement with extensive observations by Bouguer and by Pierre le Monnier (1715–99) despite the fact that the three coefficients which appear in it were determined by means of an approximate theory. Only in the particular case of horizontal refraction, and for very low altitudes, did it yield values which were much too small.[40]

Encouraged by Mayer's success, Euler took up the matter again, and developed a more general formula whose proof he communicated to Mayer on 18 March 1752 [6].[41] Mayer lost no time in testing the accuracy of this formula and found that it agreed perfectly with observations from altitudes of over 23°; on the other hand, it predicted values for the horizontal refraction which were invariably too small [7], although the error was less than in his own formula. Mayer's further efforts to account for these remaining discrepancies proved to be unsuccessful, so he contented himself with restricting his calculations to altitudes of 2° and over, which his own formula enabled him to do with sufficient accuracy. He told Euler that he took comfort from the fact that even the refraction table in the *Berlin Calendar* was not based upon a rigorous theory, but appeared to have been adjusted to suit the observations [9]. Previously, while still in Nuremberg, Mayer had drawn attention to another instance of this theoretically unjustified correction (viz., a decrease by a factor of 1:90) being applied by the computers of the 1749 almanac to the values of the Moon's semi-diameter, to make the theoretical figures which had been given for the three previous years compatible with observations.[42] He had independently derived the correction himself to compensate for a discrepancy which he had found between the values of the Moon's apparent geocentric semi-diameter based upon Euler's lunar tables[43] and the corresponding values found from his own micrometric measurements.

The precise evaluation of the Moon's semi-diameter was, of course, intimately linked to the question of the Moon's variable distance from the Earth; and Mayer was well aware of the fact that the lack of an accurate knowledge of the lunar parallax was a major drawback—indeed, in 1750 he was still regarding it as the *sole* drawback—to the application of lunar occultations in the problem of finding longitude. Thus the importance which he was now attaching to the precise evaluation of the latter quantity is understandable. He tells Euler that he had employed two methods for determining the lunar parallax but that they had yielded incompatible results, and tentatively suggests as a possible explanation of this discrepancy that the Earth's centre of gravity might not quite coincide with its geometrical centre [1]. Euler did not support this belief, and wondered whether the difficulty might not lie in the ambiguity of the definition of the parallactic angle. Should this angle, he queried, be regarded as being formed at the Moon's geometrical centre, or at the centre of its observed disc (i.e. one lunar radius nearer)? [2] In obtaining the results which he discussed in his lecture "In Parallaxin Lunae eiusdemque a terra distantiam inquisitio" (1752),[44] Mayer made a quantitative allowance for the Earth's spheroidal shape, a fact which he had hitherto been disregarding when framing the basic equations of his lunar theory but which he was now beginning to realise might be capable of producing inequalities in the Moon's motion [3].

Mayer frankly admitted to Euler [1] that it was only after he had read the latter's papers in analysis and mechanics that he had seen how to investigate the theory of the Moon's motion. He had applied the same method of approximation to the lunar theory as Euler had adopted in his prize essay to the Paris Academy "Recherches sur les irregularités du mouvement de Jupiter et de Saturne" (1748),[45] but retained the mean anomalies of the Sun and Moon instead of using their true (eccentric) anomalies as Euler had done. This had the effect of simplifying the calculation of the Moon's orbit, since the true position of the Sun was not required for the calculation of that of the Moon. He had soon discovered, however, that more reliable predictions could be obtained by reverting to the method proposed by Edmond Halley (1656–1742) for the calculation of lunar and solar eclipses, which had been based upon the Saros Cycle (viz., 223 lunations ≑ 18 years $11\frac{1}{3}$ days) and so he now felt that an empirical approach would be preferable for the determination of the lunar and solar anomalies. All that seemed to him to be required was a continuous series of observations of the Moon's celestial co-ordinates extending over a period of eighteen years—the time taken by its nodes to precess once round the ecliptic—from which lunar tables capable of yielding accurate predictions of the Moon's position could be calculated for any future time [1]. Euler agreed with Mayer that working with the mean anomalies would simplify the computation of the lunar inequalities, but pointed out that this approach had the disadvantage of complicating the formulae involving the *true* Moon–Sun distance, which was an important factor in the calculations [2].

Mayer's recognition of the need to take account of the Earth's spheroidal shape now involved him in the calculation of a large number of new inequalities, of which he retained sixteen that he considered might prove to be appreciable: one of these was an inequality which Newton himself had attributed to the variation in the eccentricity of the lunar orbit, and which Mayer suspected might have an appreciable influence upon the motion of the lunar apogee [5]. The problem of accounting theoretically for the observed value of the mean motion of this latter quantity, which was about twice as large as Newton's theory predicted it to be, had been discussed by Alexis Clairaut in his pioneering study "De l'orbite de la lune dans le système de Newton",[46] and his subsequent work had been directed towards resolving the enigma.[47] His initial reaction, like that of Euler, was to infer that the inverse-square law of gravitational attraction should be regarded only as an approximation to the truth, whereas Mayer was much more inclined to attribute discrepancies to the observations upon which the theory was based and not question the validity of the working hypothesis itself [1]. In the case of the particular discrepancy in question, it turned out that neither view was correct, since Clairaut's researches led him to recognise that its existence was due entirely to his former neglect of certain small terms in his analysis. It is, however, interesting to observe that Clairaut's (and Euler's) viewpoint was that of the theoretician, while Mayer's was that of the practical astronomer—a difference of attitude which proved to be of fundamental importance in the developments of the next few years.

Euler's unwillingness to accept the inverse-square law of gravitational attraction as being exact, stemmed from his belief that it was magnetic in origin.[48] Mayer, however, had doubts about the ability of such an assumption to account for the direct proportionality between the force of attraction and the masses of each of the bodies concerned. He thought that the number of magnetic particles in a heavenly body would not, in general, be exactly proportional to its mass; and that if no magnetic material were present in the neighbourhood of two heavenly bodies, there would be no attraction at all between them [3]. Euler dismissed this objection as being of little significance, in view of the fact that Newton's statement that attractive forces are directly proportional to the masses of the attracted bodies was merely a postulate, and had no observational justification [4]. Since, like Euler, Mayer had never been able to accept the Newtonian—though not Newton's—assumption that gravity was an essential property of matter and inherent in bodies,[49] he preferred to adopt the view of Gerhardt Müller that it was due to the pressure of the aether [3].[50] After reading Euler's "Conjectura physica circa propagationem soni ac luminis" (1750)[51] he wondered whether the agitation of aether particles arising from the propagation of light could produce a decrease in the elasticity of the aether, which could also affect this phenomenon [5]. Euler agreed that this seemed possible, but criticised Müller's assumption that the aether between two heavenly bodies has a weaker elastic force, since this was opposed to the principles of equilibrium—unless, of course, the aether were assumed to be in motion [4].

A more tangible link between these theoretical speculations and the lunar theory arose as a result of Mayer's attempts to test the validity of his belief in the indirect influence of the Earth's spheroidal shape upon the motion of the lunar apogee. In the course of this investigation, Mayer compared contemporary observations with those made by John Flamsteed (1646–1718) and other astronomers during the previous century, and found that the former yielded a smaller value for this quantity than did the latter. At the same time, he discovered that the Moon's mean motion in longitude was increasing slowly but steadily; in other words, that there was a secular acceleration in the Moon's mean motion [10]. The existence of this acceleration (which Mayer first fixed at 7″, but later at 9″, per century), had previously been detected by Edmond Halley and acknowledged by Newton. In extending his investigations to the ancient observations of Hipparchus and Ptolemy, however, Mayer also detected an error of $1\frac{1}{4}°$ in Ptolemy's estimate of the time of an equinox, and thereby accounted for a phenomenon which Euler had just been ascribing to an apparent secular acceleration in the *Sun's* mean motion or retardation of the Earth's own motion resulting from the resistance of an aether [17]. Thus Mayer had effectively revealed the irrelevance of the aether hypothesis in the explanation of discrepancies between the theory and observation of the Sun's motion.

Mayer's researches on the length of the tropical year led him to believe that there was some observational justification for believing that the eccentricity of the Earth's orbit was gradually diminishing, and he was aware that this was indeed likely to give rise to a retardation in the Earth's motion through the combined attraction of Jupiter and Venus, just as the observed retardation in the motion of Saturn had been shown by Euler to be linked to a decrease in the eccentricity of its orbit. Yet there was no evidence of an appreciable secular acceleration in the length of the year such as Jacques Cassini's results in his *Élémens d'Astronomie* (Paris, 1740) seemed to suggest. The fact that no retardation in the Earth's motion had ever been observed, could be explained by supposing that such an effect really did exist (for the above-mentioned reason) but that it was exactly compensated by the resistance of the hypothetical aether which at the same time would produce the "very marked acceleration in the Moon's motion" [18]. However, Mayer preferred to believe that the (remaining) lunar inequalities which the structure of his theory enabled him to evaluate without too much difficulty resulted from the influence of the Sun's variable gravitational attraction and, although he was unable to derive the theoretical implications of this assertion, his arguments were sufficient to convince Euler that he had been wrong in postulating the secular acceleration of the Sun. Nevertheless, Euler felt justified in maintaining that the *observed* secular acceleration of the Moon's mean motion could result from a periodic inequality with a period of many centuries [19]. The aether, however, no longer seemed to be necessary to the explanation of discrepancies between the theory and observations of the Moon's motion.

Meanwhile, Clairaut's success in accounting theoretically for the observed mean motion of the lunar apogee had encouraged him to enter the competition sponsored by the St Petersburg Academy of Sciences in 1751, the subject of which was:

> To demonstrate whether all the inequalities observed in lunar motion are in accordance with Newtonian theory—and if they are not, to demonstrate the true theory behind all these inequalities, so that the exact position of the Moon at any time can be computed with the aid of this theory.[52]

The title of his treatise, "Théorie de la lune déduite d'un seul principe de l'attraction réciproquement proportionelle aux quarrés des distances", reveals his acceptance of the Newtonian hypothesis as a sufficient basis for the explanation of the observed motion of the Moon. Euler, whose own thorough re-examination of the lunar theory had just led him independently to the same conclusion was quick to accept Clairaut's argument and (in his capacity as an official adjudicator of this contest) recommended this essay as being most worthy of the honour which the St Petersburg Academy of Sciences duly bestowed upon it. Through Euler's influence, it was published at the expense of the Berlin Academy in 1752.

Clairaut's theory was founded upon a polar system of co-ordinates, into which he introduced perturbation terms using the mean anomaly of the perturbed body (viz., the Moon) as the independent variable. This procedure, however, possessed the disadvantages of giving rise to complicated expressions and necessitating laborious investigations on the convergency of the equations for the perturbing force; on the other hand, a convergent expansion for the true lunar anomaly was obtained by substituting, as a first approximation, the values of the observed eccentricity of the lunar orbit and the motion of its apsides into the equation for a variable ellipse. The major difficulty was that the small dividing factors in the integration gave rise to large coefficients in the perturbation terms and yielded secular equations dependent upon the ratio of the mean motions of the disturbing and disturbed bodies (viz., the Sun and the Moon respectively). Clairaut followed Newton's method in determining the variations of the node and nutation.[53] He compared his theoretical predictions of the Moon's celestial co-ordinates with 99 observations referring to a period of about ten years extracted from the catalogues of Jacques Cassini (1677–1786) and Jean Dominique Maraldi (1709–88) and reduced by the Abbé Nicolas de la Caille (1713–62). The results were, however, rather unsatisfactory: the celestial longitudes differed from between $+6'$ to $-5\frac{1}{2}'$ with a mean of over $2'$, while the corresponding latitudes differed from over $+2'$ to $-3'$ with a mean of $45''$.[54] Jean d'Alembert's *Théorie de la lune* (Paris, 1754) appeared to possess a similar degree of reliability, despite the fact that it incorporated the improvement of deducing the motion of the Moon's apsides theoretically instead of regarding this as a (superfluous) element of observation.

Euler's own *Theoria motuum Lunae* (Berlin, 1753)[55] differs fundamentally from Clairaut's in being based upon a Cartesian co-ordinate system; the changes in the polar co-ordinates and perturbations in the nodes and nutation being deduced purely analytically from the three equations for the mutually perpendicular components of the accelerating force. He introduced a second-order term in Newton's law of gravity—his reason for this has already been indicated—but found it to be negligible (which is why he had supported Clairaut's conclusion). He simplified the problem of integration by developing a method in which the six orbital elements[56] were regarded as variable and using six first-order differential equations instead of three second-order ones. A co-ordinate system rotating with the mean speed of the Moon, which he employed, had the advantage that the co-ordinates themselves were small in comparison with the perturbations. He also introduced in this context a method of undetermined coefficients, which the famous French mathematician Joseph Louis de Lagrange (1736–1813) later found to be an important analytical tool in his investigation of the stability of the solar system. The inadequacy of Euler's theory lay in its failure to take into account the secular acceleration of the Moon's mean motion, the divergences of both the Moon and the Earth from sphericity, and planetary perturbations (particularly from Venus)[57]; and this is why its predictions were found to err by about $\pm 5'$.

Mayer's "Novae tabulae motuum solis et lunae" (1753),[58] which were the outcome of the researches outlined above, appeared in print at about the same time as Euler's lunar theory. Unfortunately, Mayer gave no indication in this publication of how he had arrived at his formulae or what method of integration he had used, yet was claiming for them an accuracy of within 2′ which was significantly in excess of that found by both Clairaut and Euler. Thus one can forgive d'Alembert for being sceptical of Mayer's achievement [21, 22]. At the end of his preliminary discussion to his tables, the latter remarks that the situation had now been attained where the accuracy of the observations could not match that of the theory; this was a belief which was fully confirmed by the results of his further comparisons with observed data on the Moon's position. In particular, an uninterrupted series of 139 lunar positions observed at Greenwich during the years 1743–45 by the Astronomer Royal James Bradley—transmitted to Mayer via Euler and Christopher Schumacher (1704–68)—agreed to within 1′ with his own predictions; and so did similar observations published by Richard Dunthorne (1711–75) although they often exhibited differences of between 3′ and 4′ with the earlier observations of John Flamsteed. An independent verification of this exceptional accuracy in his theory—at any rate, for the particular case of the Moon's conjunctions and oppositions with the Sun—was that its differences with respect to the Moon's celestial position derived from a systematic study of lunar and solar eclipses observed after the invention of the astronomical telescope and pendulum clock never exceeded 1′ 10″; and Mayer was of the opinion that these discrepancies could have been largely due to the variation in the eccentricity of the Earth's orbit [12]. Bradley's observations, and a catalogue of lunar eclipses arranged in order of their Saros cycles, are appended to Mayer's "Tabularium lunarium in commentt. S.R. Tom. II contentarum usus in investiganda longitudine maris" (1754),[59] in which he stresses the usefulness of his tables as a basis for the discovery of longitude at sea. An uncertainty of 1′ in the Moon's position in the sky is approximately equivalent to an error of $\frac{1}{2}°$ (or 30 nautical miles) in terrestrial longitude;[60] and this was reliable enough for the purposes of oceanic navigation.

In fact, the accuracy which Mayer had attained was sufficient to entitle him to one of three bounties offered by the British Parliament under the terms of the Act 12 Anne (1714) "for Providing a Publick Reward for such Person or Persons as shall Discover the Longitude at Sea".[61] But Mayer was well aware that the method of lunar occultations, though adequate for establishing a single fixed longitude difference between two land-based observers, was unsuitable—essentially on account of the infrequency of this phenomenon—as the basis of a *general* means of finding the variable longitude difference between a ship at sea and an observer situated on some standard meridian. A further difficulty was the practical one of making the necessary measurements of the angular distances between the Moon and the star before and after the occurrence of this phenomenon (as was Mayer's practice)—or, alternatively, of accurately estimating the times of immersion and emersion—on a ship which would be rocking and rolling on the high seas [22]. Mayer's tables ensured that the mariner could be provided in advance with a list of accurate times at which certain occultations were predicted to occur, but this was not a sufficient condition for the method being generally "useful and practicable" as the 1714 Act required.

It was as a result of his consideration of the two independent observational means of testing the accuracy of his lunar theory (viz., lunar occultations, and lunar and solar eclipses) that Mayer, about the end of 1753, came to recognise the fundamentally important fact that a principal cause of the discrepancies of 5′ and over found in the lunar theories of Clairaut, Euler and d'Alembert was that errors in observed star positions were being reflected in the observed celestial co-ordinates of the Moon. Such errors would have no effect upon the values of the coefficients in the lunar inequalities, but they would clearly affect the value of the Moon's mean motion deduced from the measured time interval between the star's immersion and emersion behind the lunar disc and could easily be corrected with the aid of solar eclipse observations. Working on this principle, he made use of his own observations of the occultations of the bright stars Aldebaran, Spica, Regulus, Antares, the Pleiades, etc., to compute over a hundred values of the Moon's longitude to within only 5″ to 10″. He had also managed to double the internal accuracy of his tables by taking account of a larger number of terms in the infinite series, so he was now in a position to judge the validity of the theoretical basis upon which these tables rested. Converse calculations, in which he made use of those results to deduce the positions of other stars, revealed errors in contemporary star catalogues which (if ignored) would have produced inaccuracies of several minutes of arc in the Moon's coordinates derived directly from observations of these stars' occultations. The same errors had induced able astronomers such as Pierre Le Monnier (1715–99) to believe that the star Regulus was exhibiting a proper motion which in reality did not exist [20].[62]

Thus it is easy to understand why Mayer's subsequent astronomical researches were devoted to: the elimination of instrumental errors in a mural quadrant employed for the measurement of star positions (1756); a simple and accurate method of computing solar eclipses (1757); the construction of a catalogue of zodiacal stars (1759); an investigation of the proper motion of the stars (1760). These are, in fact, the respective topics of four of the six treatises edited by Georg Christoph Lichtenberg in his *Opera inedita Tobiae Mayeri I* (Göttingen, 1775), the only volume of Mayer's unpublished works yet to have been printed.[63] A fifth treatise is devoted to the problem of accurately defining the variations of a thermometer (1755),[64] which Mayer claims should be solved in the manner which had proved to be so successful in astronomy; namely, isolating a major periodicity, examining the residuals for a second-order periodicity, etc., until the observations had been analysed into a series of independent periodic functions each characterised by a mean value and its variation about that mean. The relationship of this treatise to his investigations in the lunar theory is obvious. The remaining treatise is concerned with colour mixing (1758),[65] a hobby to which Mayer seems to have returned when circumstances prevented him from continuing his astronomical observations: one could, however, relate this interest to the problem of preparing accurate reproductions of maps or to a latent desire to reduce all natural phenomena to quantitative mathematical terms. Lichtenberg appears to have been totally unaware of any inner connections among these treatises—at any rate, he fails to point them out and his criteria of selection were not in the least concerned with this kind of consideration. Of greatest interest to us here, of course, is their common link with Mayer's studies in lunar theory.

The only person among all Mayer's contemporaries who could really appreciate the value of his contributions to this subject was Euler himself. It was in consequence of his open assertion after the publication of Mayer's lunar tables—"the most admirable masterpiece in theoretical astronomy" [19]—that no astronomer had improved, or would improve, the lunar theory more than Mayer, that he was instructed by the President of the Berlin Academy (Pierre de Maupertuis) and its members to find out whether Mayer would be interested in taking up an appointment in Berlin. Euler duly communicated this request in his letter of 24 April 1753, begging Mayer to treat the matter confidentially [11]. The latter's reaction, as might be expected, was clearly one of pleasure; and although he felt that little financial advantage would accrue from such a move, he valued the possibility of having contact with great scholars more highly than all other forms of improvement [12]. The precise terms of the offer could not, however, be settled since Maupertuis was not then resident in Berlin and Mayer expected an appreciably larger remuneration than the Academy's treasury could bear.

While this matter was still under discussion, Euler was formally requested by the Secretary of the St Petersburg Academy (G. F. Müller) to offer Mayer a professorship there.[66] The salary was to be considerably greater, but Euler advised Mayer to await Maupertuis' return after which attempts could be made to improve the offer from the Berlin Academy [21]. Maupertuis finally managed to persuade Frederick the Great, who was then keeping a tight control of that Academy's purse strings, to grant to Mayer the salary which he wanted, a sum for travelling expenses and free accommodation thereafter in Berlin. His sole duty would be to attend regularly the weekly meetings of the local Academy and lecture occasionally on his research; while the Berlin Observatory would be placed entirely at his disposal, in the hope that he would continue the annual publication of the *Berlin Calendar* in the stead of

J. Kiess (1713–81) who was leaving to take up a post at Tübingen University [24]. Mayer gave Euler his assurance that not only would he willingly construct this calendar in a more popular and convenient manner, but would include lunar ephemerides based upon his tables whose accuracy he could now guarantee to within 30″. He would make it one of his first duties in Berlin, to have these tables printed together with some relevant remarks and a justification of their validity [26].

Mayer duly received a friendly letter from Maupertuis himself, confirming the financial offer,[67] and only four days later Euler forwarded a letter to Mayer from Müller[68] containing the formal proposals from the St Petersburg Academy [25]. He had no hesitation in rejecting the latter offer [27], in accordance with Euler's advice, and wrote to Gerlach Adolph von Munchhausen (1688–1770) whom George II had entrusted with the control of his Hanoverian affairs, informing him of the Berlin Academy's proposal and requesting his resignation from the Georg-August Academy (which George II himself had founded in 1737).[69] Contrary to his expectation, however, this request was not granted; instead, he was asked to stipulate the conditions under which he would prefer to remain in Göttingen and—much to his surprise—was granted them. He was to be given a salary in excess of what he would otherwise have been receiving in Berlin, and the sole directorship of the Göttingen Observatory (then nearing completion) which he had been sharing since 1751 with Professor U. von Segner (1704–77).[70] Thus he made up his mind to resist the lure of the big city and to remain in Göttingen, and confided to Euler that he had been thought of in Hanover more highly than he deserved [28].

It was only six days after communicating this information to Euler that Mayer read his "Methodus Longitudinum Promota" to a meeting of the Göttingen Scientific Society.[71] This was a manuscript outlining the precepts by which he would find the longitude at sea with the aid of his lunar tables; his intention when composing it was to transmit it to London in support of a bid which Euler had already been encouraging him to make [21] for one of the longitude prizes offered by the British Parliament.[72] The advantage of the method was that it involved a single observation only, of the distance between the Moon and a zodiacal star. The disadvantage was that the longitude had to be assumed as known so that this lunar distance could be interpolated from his tables and then compared with the observed distance. Any disagreement between the measured and calculated values could then be substituted into another relation and the resulting correction to the assumed longitude thus found by a "Regula falsi" [26]. On the same occasion, Mayer described a new instrument based upon the application of his repeating principle to a reflecting octant of the type designed by John Hadley[73] which he considered would be preferable to the common sea-octant for making the necessary measurement of the lunar distance. As soon as he heard through Johann David Michaelis, Secretary for Hanoverian Affairs in Göttingen, that informed opinion in England favoured his chance of success,[74] he transmitted the above-mentioned manuscript and a copy of his lunar tables to the First Lord of the Admiralty (Lord Anson) via the official governmental channels: they were in his hands by 20 January 1755.[75] Lord Anson was, at the same time, the President of the Board of Longitude, the body of (originally 22) commissioners nominated in the 1714 Act to administer the Parliamentary awards and to adjudicate all proposals relating to the problem of longitude determination at sea.[76]

Before the end of 1755, in compliance with the request of the Secretary of State for Hanoverian Affairs in London (William Philip Best),[77] Mayer had forwarded another manuscript in which he described the principles upon which the calculations in his lunar tables had been founded; new manuscript lunar tables which took account of the important improvements that he had made during the course of the past two years; and a wooden model of the repeating circle which he had described in his lecture, ruled by himself. Contrary to Mayer's expectation, sea trials by Captain John Campbell of a smaller-scale brass model of this instrument showed that it possessed no appreciable advantages over the common Hadley octant. On the other hand, it was as a direct result of these experiments that Campbell conceived the idea of combining the respective merits of both Hadley's and Mayer's designs in his nautical sextant.[78] The application of Mayer's repeating principle was soon rendered obsolete by Jesse Ramsden's invention of a machine for accurately ruling the limbs of the conventional octants and sextants,[79] so Mayer's Circle never became popular for use at sea. Modifications to his design made by Jean Charles Borda (1733–99), although of little advantage in this respect, nevertheless had the effect of rendering it suitable for use before the end of the century in the French measurements along an arc of the meridian to establish the length of the standard metre.[80]

The completion of the Göttingen Observatory in 1754 and the installation there, in the autumn of the following year, of an excellent 6 ft radius mural quadrant specially constructed by the talented London instrument-maker John Bird (1709–86)[81] placed Mayer in a better position to make his own contributions to positional astronomy. With the aid of this new quadrant and a pendulum clock, he made a large number of observations of the zodiacal stars from 1756 to 1761 inclusive[82] despite frequent interruptions due to the military occupation of Göttingen by French troops. Not only did the Seven Years' War have an adverse effect on Mayer's astronomical observations, in making it unsafe for him to go to his observatory at night,[83] but it was also a significant factor in postponing consideration by the Board of Longitude of Mayer's claim for a longitude prize. Due to Britain's involvement in that war, and due to the Board's long-standing commitment to encouraging John Harrison (1693–1776)[84] to construct a marine timekeeper that would make practicable an alternative method for finding longitude at sea,[85] no decision was taken on the question of an award to Mayer prior to his untimely death on 20 February 1762. Shortly before he died, however, he received lunar distance observations made at sea with the aid of a Hadley octant by a former Danish student at Göttingen (Carsten Niebuhr) whom he had specially trained to carry out such a task. These pleased him so much that he asked his wife to send them after his death to England, along with a copy of his solar and lunar tables containing numerous minor improvements that he had gradually been making to them during the previous seven years.[86] Despite his reservations concerning the impartiality of the commissioners of longitude to claims made by foreigners, he nevertheless entertained the hope that they would now recognise the practical reliability of his method, and see fit to award the smallest of the three bounties (£10,000) to his wife and—at that time—four surviving children.[87]

The Board of Longitude's recommendation to the British Parliament in February 1765 to award a sum "not exceeding £5000" to Mayer's heirs, which that Parliament decided three months later to fix at £3000,[88] reflects the fact that the commissioners were regarding Mayer's "last manuscript lunar tables"[89] as a necessary *but not sufficient* basis for finding longitude at sea. Auxiliary tables, further instructions, worked examples, etc., had still to be supplied in order to reduce the complexity of the calculations involved before the method of lunar distances which Mayer was championing became both useful *and practicable* at sea, in accordance with the terms of the 1714 Act. This was achieved by the Rev. Nevil Maskelyne (1732–1811) through the publication of *The British Mariner's Guide* (London, 1763), the *Nautical Almanac for 1767* (London, 1766) and for subsequent years,[90] and his "Requisite Tables" (London, 1767)[91] for "clearing" the effects of refraction and parallax from the observed lunar distances. Since, however, no fundamentally new principle was involved in these labours, the question of Maskelyne's own entitlement to a major Parliamentary award never arose. Besides these works, Maskelyne was responsible, in his capacity as Astronomer Royal, for supervising the printing and publishing of the manuscript that Mayer had composed and transmitted in 1755 containing his derivation of the formulae on which his tables had been based, viz., *Theoria lunae juxta systema Newtonianum* (London, 1767). He also thoroughly edited the "New and Correct Tables of the Motions of the Sun and Moon" which had been sent to London by Mayer's widow in the summer of 1763, and had them published both in Latin and in English in 1770.[92] This latter work contains Mayer's "Methodus longitudinum promota", which appears after the preface, and a diagram of Mayer's Repeating Circle. With the publication of this work, all the essential requirements for the fruitful application of the method of lunars were fulfilled, and this became the first really reliable means of longitude determination at sea which mariners all over the world were trained to understand and adopt.

The outstanding problem in the lunar theory itself which had still to be satisfactorily resolved was that of giving a physical explanation of the Moon's secular acceleration, and the Paris Academy in 1770 offered a prize for its solution. Euler won this prize, yet his conclusion that "the secular inequality in the Moon's mean motion cannot be produced by the forces of gravitation",[93] might be taken as an indication that he was persisting in his belief in a resisting medium. That this was indeed the case is verified by his solution to the same problem two years later, on which occasion he shared another of the Paris Academy's prizes with Lagrange.[94] Lagrange himself won a further prize in 1774 for an investigation of the gravitational effects of the Earth's and Moon's spheroidal shapes, but failed to find a secular equation: he did show, however, that the attraction of planets and satellites could not be the cause of the phenomenon and his critical discussion of the observations upon which the alleged acceleration was founded even made him doubt the reality of its existence. Laplace, on the other hand, considered the effect to be fully established, but did not accept Euler's explanation since there was no independent proof of the existence of an aetherial fluid.[95] A significant improvement of a different nature was made by Dr Bradley's former assistant Charles Mason who, with the help of the former's own lunar observations (which, like those of Mayer and La Caille, refer to the period from 1750 to 1760), succeeded in reducing the errors in Mayer's tables still further: nevertheless, he still found differences of 19″ or more in values of the Moon's celestial longitude although he quotes his figures to the nearest 0·1″.[96] Mason's improved "Tables of 1780" incorporate for the first time the effects of planetary precession upon the celestial co-ordinates of the stars, a phenomenon which Mayer had neglected.

In 1787, Laplace triumphantly announced that he had discovered the cause of the phenomenon in the gradual diminution of the mean action of the Sun.[97] The actual amount of the predicted acceleration depended upon the variation of the eccentricity of the Earth's orbit, which required a knowledge of the planetary masses; fortunately, the effect was greatest in the case of Jupiter whose mass was well known from observations of its satellites. The same cause was also responsible for a secular retardation in the motion of the Moon's perigee and nodes. Lagrange, by applying a former analysis of the secular variations in the motions of Jupiter and Saturn to the case of the Moon, confirmed Laplace's conclusion.[98] By the beginning of the nineteenth century, the discovery of these and other new equations in the lunar theory and the substitution of certain theoretical terms in place of their empirical equivalents (e.g. the inequality in celestial longitude due to the Earth's spheroidal shape), had decreased the mean error in the predicted values of both of the Moon's co-ordinates to about 8″.

Although further advances in the mathematical structure of the theory led Denis Poisson (1781–1840) to discover that Laplace had not succeeded in accounting quantitatively for the Moon's secular acceleration,[99] the predictive ability has never been substantially improved. The effect of tidal friction and other geophysical and meteorological factors, are now known to give rise to appreciable irregularities in the Moon's motion and in the speed of the Earth's axial rotation, whose existence Mayer was prepared to postulate [18] although he had no idea of their causes. Thus computations of lunar distances made to within 0·1′ (or 6″) would have been amply sufficient for the purposes of eighteenth-century navigation, and had this been appreciated the methods of lunars could have been greatly simplified.[100] But can we blame Maskelyne and his contemporaries for striving to achieve refinements in observational method and technique, and for not suspecting the intrinsic limitations to the power of analysis? It is only with the hindsight gained from a century and a half of intensive research in the astronomical and geophysical sciences, that we are now able to recognise that celestial motions can never be described in strictly geometric terms nor controlled by a single universal law. Ironically, we differ from Newton himself not in our conclusion, but only in the premises on which it is based!

REFERENCES

1. This was first published in F. X. von Zach, *Monatliche Correspondenz zur Beförderung der Erd-und Himmels-Kunde*, ix (1804), 415–32; and reprinted as part of the introduction (viz., pp. xxxv–li) to J. F. Benzenberg, *Erstlinge von Tobias Mayer* (Dusseldorf, 1812). A manuscript copy dated 13 August 1840 made from the original by a Dr Schweiger in Göttingen shortly after this was sold at a local book auction to the Ducal library in Gotha, and is preserved in the Göttingen University library under Cod. MS. Hist. lit. 18.
2. The most reliable source of information on Mayer's early years in Esslingen is P. Eberhardt, "Urkundliche Beiträge zu der Jugendgeschichte des Astronomen Johann Tobias Mayer", *Literarische Beiträge des Staatsanzeigers* (Stuttgart, 1908), pp. 177–87. This article was later printed with several minor amendments in P. Eberhardt, *Aus Alt-Esslingen* (Esslingen, 1924), pp. 207–24.
3. J. M. Franz, *Homännischer Bericht von Verfertigung grosser Weltkugeln* (Nuremberg, 1746).
4. The titles of these maps (Nos 6 and 7) in the atlas, are:
   (i) "Belgii universi seu Inferioris Germaniae, quam XVII. Provinciae Austriaco, Gallico & Batavo sceptio parentes constituunt, nova tabula geographica, a Tob. Majero Math. Cult. ad leges legitimae delineationes revocato 1747."
   (ii) "S.R.I. Circulus Austriacus, compositus ex Ducatu Stiriae, Duc. Carinthiae, Duc. Carnioliae, Comit, Tyrolensi & ditionibus in Sueviae anterioribus Austriacis cum suis confiniis a Tob. Majero Math. Cult. legitime designatur. A. 1747."
5. "Sammlungen verschiedener fremden und eignen Beobachtungen von Länge und Breite . . ." and "Collectanea geographica et mathematica", listed as Cod. MS. T. Mayer 10 and 15_{11} respectively in the *Verzeichniss der Handschriften im Preussischen Staate*, I Hannover 3 Göttingen, 3 (Berlin, 1894), pp. 154–8.
6. *Ibid.*, Mayer 15_{11}.
7. The aims of this society are described by its founder Johann Michael Franz, in his *Homannische Vorschläge von den nöthigen Verbesserungen der Weltbeschreibungswissenschaft und einer disfals bey der Homannischen Handlung zu errichtenden neuen Academie* (Nuremberg, 1747).
8. Tab. XXI "Vorstellung der in der Nacht zwischen den 8. und 9. Aug. 1748 vorfallenden partialen Mond Finsternis . . ." (1748), in J. G. Doppelmaier's *Atlas Coelestis* (Nuremberg, 1752).
9. The German title of this collection of ephemerides, etc., is: "Historisch und geographischer Calendar . . . unter Genenmhaltung der von Sr. Konigl. Maj. in Preussen in Dero Residenz Berlin gestifteten Akademie der Wissenschaften".
10. Five of these are listed by Franz in *op. cit.*, 7.
11. J. Hevelius, *Selenographia* (Danzig, 1647).

12. G. B. Riccioli, *Almagestum Novum Astronomiam Veterem Novamque Complectens Observationibus Aliorum* (Bologna, 1651). Cf. Figs. 1–6.
13. Mayer's unpublished "Collectanea mathematica 1747–53" (*op. cit.*, 5, Mayer 12) contains extracts of observations of Jupiter's satellites, solar and lunar eclipses, lunar occultations, etc., which he had collected from various sources.
14. Mayer describes his micrometer in the *Kosmographische Nachrichten u. Sammlungen auf das Jahr 1748* (Nuremberg, 1750), pp. 1–11.
15. "Beobachtungen einiger Zusammenkunften des Monds mit Fixsternen im Jahr 1747 und 1748 zu Nürnberg in dem Homännischen Hause angestellet", *ibid.*, pp. 41–51.
16. *Ibid.*, pp. 397–419.
17. *Histoire de L'Académie Royale des Sciences et Belles-Lettres. Année MDCCLII* (1754), 262–282.
18. C. Mylius, *Gedanken über die Atmosphäre des Monds* (Hamburg, 1746).
19. C. Mylius, "Gedanken über des Herrn Prof. Mayers in Göttingen Beweis, daß der Mond keinen Luftkreis habe", *Physikalische Belustigungen*, i (1751), 288–302.
20. The article by Euler to which Mylius refers is: "Sur l'atmosphère de la lune prouvée par la dernière Éclipse annulaire du Soleil", *loc. cit.*, 17, *Année MDCCXLVIII* (1750), 103–121.
21. *Loc. cit.*, 14, pp. 52–183. This is also the source of the information contained in the previous paragraph.
22. *Bericht von den Mondskugeln, welche bey der kosmographischen Gesellschaft in Nürnberg, aus neuen Beobachtungen verfertiget werden durch Tobias Mayern. Mitgliede derselben Gesellschaft. Zu finden in der Homannischen Officin* (Nuremberg, 1750). The authorship of this 24-page pamphlet is usually ascribed to Mayer, but it is just as likely to have been written by Franz.
23. One reason was Mayer's departure from Nuremberg early in 1751; another, the bankruptcy of the Homann firm due to the over-ambitious projects launched by Franz. For further information on this point, the reader should consult the articles by S. Ruge, "Aus der Sturm- und Drang-Periode der Geographie", *Zeitschrift für Wissenschaftliche Geographie*, v (1885), 249–60; and C. Sandler "Die homännischen Erben", *ibid.*, vii (1890), 332–55 and 418–48.
24. These were drawn by the director of that school, Georg Martin Preissler (1700–54). See *loc. cit.*, 5, Mayer 15_{52}.
25. J. Russell, *A Description of the Selenographia; an apparatus for exhibiting the phenomena of the Moon* (London, 1797). In this work, Russell uses Mayer's principle of the micrometric determination of numerous markings.
26. *Personalakte Tobias Mayer Nummer 4/Vb 18*, Dekanate u. Universität Archiv, Göttingen: letter from Gerhardt Adolph von Munchhausen in Hanover to Tobias Mayer, Mathematician in Nuremberg, 26 November 1750. This letter is a copy of the original, which was initially in the possession of Hofrath Scheidt, Göttingen.
27. An enquiry addressed to the *Evang. Stadtpfarramt* in Neckarbischofsheim elicited the following piece of biographical information: Maria Victoria Gnüge, born on 30 July 1731, was the youngest of a family of six children. Her father, Johann Christoph Gnüge (originally from Gotha, Thüringen) was a minister in that town from 1704 until his death in 1747. This evidence conflicts with that extracted from the "Kirchenbuch der Be-

grabenen" (Albani Gemeine) in the *Evangelische Kirchenbuchamt*, Göttingen, where it is stated in an entry for 22 February 1780 that "Frau Maria Victoria Mayer gebohrne Gnüg, seligen H. Profess: Mayers nachgel: Wittwe" was 58 years 11 months 15 days old when she died on 14 February 1780. This would mean that she was born on 7 March 1723. We regard the former source as the more reliable.

28. "Germania atque in ea locorum principaliorum mappa critica ex latitudinum observationibus, quas hactenus colligere licuit, omnibus mappis specialibus compluribus; itinerariis antiquis Antonini, Augustano et Hiemosolymitano, adhibita circumspectione ac saniori crisi concinnata simulque cum aliorum Geographorum mappis comparata a Tob. Mayero, Societatis cosmographicae sodali, impensis Homanniorum Heredum." For brevity, this is usually referred to as Mayer's "mappa critica".
29. "Johann Michael Franz Vorschläge, wie die Erdbeschreibung in Absicht Deutschlands zu verbessern sey," *loc. cit.*, 14, pp. 348–96 (cf. p. 352).
30. Translated from *loc. cit.*, 5, Mayer 15_{11}.
31. Mayer's lecture on the improvements which he had made to this instrument, entitled "Instrumenti goniometrici, quod astrolabium vocant, structura emendatior", was delivered to the Göttingen Scientific Society on the 8 September 1759. An abstract of its content is given in the *Göttingische Anzeigen von gelehrten Sachen* (1759), pp. 993–5. An incomplete manuscript, containing sketches of Mayer's astrolabe, is contained among his unpublished papers (*loc. cit.*, 5, Mayer 15_{23}).
32. T. Mayer, "Nova methodus perficiendi instrumenta geometrica et novum instrumentum goniometricum", *Commentarii Societatis Regiae Scientiarum Gottingensis*, ii (1753), 325–36. A German translation of this article is given by Benzenberg (*op. cit.*, 1).
33. T. Mayer, "Experimenta circa visus aciem", *ibid.*, iv (1755), 120–35.
34. *Novi Commentarii Academiae Scientiarum Imperialis Petropolitanae*, i (1750), 67–105.
35. *Opuscula* (*varii argumenti*), ii (1750), 1–22.
36. *Ibid.*, i (1746), 169–244.
37. T. Mayer, *De refractionibus objectorum terrestrium* (Göttingen, 1751).
38. This was published in 1749. Mayer is probably referring here to notes devoted to various meteorological phenomena, which he ascribes to Euler himself.
39. P. Bouguer, "Sur les réfractions astronomiques dans la zone torride", *Histoire de L'Académie Royale des Sciences de Paris. Année MDCCXXXIX* (1741); *Mémoires*, 407–23.
40. R. Wolf, *Handbuch der Astronomie ihrer Geschichte und Litteratur*, ii (Zürich, 1890), pp. 269–70, compares the values of the refraction calculated from the respective formulae of Bradley, Mayer, and La Caille at five-degree intervals from the zenith to the horizon, with those actually observed. This confirms that the errors in Mayer's formula become significant only between an altitude of 5° and the horizon, and that the maximum error (for the horizontal parallax) was just over 4′.
41. Two articles on this subject published by Euler after this date, are:

(a) "De la réfraction de la lumière en passant par l'atmosphère selon les divers degrés tant de la chaleur que de l'élasticité de l'air", *loc. cit.*, 17, *Année MDCCLIV* (1756), 131–72.

(b) "Sur l'effet de la réfraction dans les observations terrestres", *Acta Academiae Scientiarum Imperialis Petropolitanae. Anno MDCCLXXVII*, ii (1780), 129–58.

42. This is discussed by Mayer in his "Astronomische Beobachtung der grossen Sonnenfinsterniß J.J. 1748. den 25. Julius zu Nürnberg in dem Homannischen Hause angestellet. Mit nöthigen Anmerkungen", *loc. cit.*, 14, pp. 11–40.
43. That is, those contained in *loc. cit.*, 35, i (1746), 149–68, which were based on the London (i.e. Greenwich) meridian.
44. Read to the Göttingen Scientific Society on 8 April 1752 and published in *loc.cit.*, 32, 159–82.
45. See for example: *Recueil des pieces qui ont remporté les prix de l'Académie Royale des Sciences, depuis leur fondation*, vii (1769), 1–84. In the "Avertissement" at the beginning of this volume, it is stated that the prize was awarded to Euler for this treatise in 1748, and that it was published in 1749.
46. *Loc. cit.*, 39, *Année MDCCXLIII* (1746); Mémoires, 17–32.
47. A. C. Clairaut, "Du sytème du monde dans les principes de la gravitation universelle," *ibid.*, *Année MDCCXLV* (1749); Mémoires, 329-90 and "De l'orbite de la Lune en ne négligeant pas les quarrés des quantités de même ordre que les forces perturbatrices", *ibid.*, *Année MDCCXLVIII* (1752), 421–40.
48. L. Euler, "Dissertatio de magnete ab Illust. Academia Regia Paris. Scient. praemio condectorata A.1744", *loc. cit.*, 35, iii (1751), 1–53.
49. Isaac Newton, in a letter to Richard Bentley dated 17 January 1693, writes: "You sometimes speak of Gravity as essential and inherent to Matter. Pray do not ascribe that Notion to me; for the Cause of Gravity is what I do not pretend to know, and therefore would take more time to consider of it." Unfortunately, as a result of a remark by Roger Cotes in the preface to the second edition of Newton's *Philosophiae Naturalis Principia Mathematica* (London, 1713) that "gravity is to be found in all bodies universally", the notion *was* erroneously ascribed to him.
50. G. A. Müller, "Untersuchung der wahren Ursache von Newtons allgemeinen Schwere; wie auch bewegenden Kräfte der Körper" (Weimar, 1746).
51. *Op. cit.*, 35.
52. Yu. Kh. Kopelevich, "The Petersburg Astronomy Contest in 1751," *Soviet Astr.—AJ*, ix (1966), 653–60 (cf. p. 653). This is a translation from the Russian original, which appeared in *Astr. Zh.*, xlii (1965), 845–53.
53. The foregoing statements in this paragraph have been based upon the analysis given by Leo Courvoisier, in his prefatory remarks to his edition of Euler's "Theoria Motuum Lunae nova methodo pertractata", *Leonhardi Euleri Opera Omnia, Ser. Secunda* xxii (Lausanne, 1958), p. xxvii.
54. A. Clairaut, *Théorie de la lune déduite d'un seul principe de l'attraction reciproquement proportionelle aux quarrés des distances* (Berlin, 1752).
55. The full title of Euler's treatise is: "Theoria motuum Lunae exhibens omnes ejus inaequalitates. In addimento hoc argumentum aliter tractatur simulque ostenditur quemadmodum motus Lunae cum omnibus inaequalitatibus innumeris aliis modis repraesentari et ad calculum revocari possit." His alternative treatment consisted in illustrating how the Moon's

motion could be represented and calculated by means of dimensionless parameters.

56. These are: (i) the major axis of the orbit; (ii) the eccentricity of the orbit; (iii) the direction of the line of apsides; (iv) the inclination of the orbit with respect to the ecliptic; (v) the direction of the line of nodes; (vi) the longitude of the epoch.
57. The foregoing statements in this paragraph have been based on L. Courvoisier, "Zur Vorgeschichte der Eulerschen Mondtheorie", *op. cit.*, 53, pp. xxvii–xxix.
58. *Loc. cit.*, 32, pp. 383–430.
59. *Loc. cit.*, 32, iii (1754), 375–96.
60. Since the Moon's mean daily motion with respect to the stars is about 13°, its mean hourly motion is roughly $\frac{1}{2}$°. Thus, an error of 1′ in the Moon's position corresponds to an error of 2 minutes, during which time the Earth rotates $\frac{1}{2}$° about its axis.
61. Further details are given in: E. G. Forbes, "Tobias Mayer's Lunar Tables", *Ann. Sci.*, xxii (1966), 105–16.
62. P. C. Lemonnier, *Institutions astronomiques* (Paris, 1746).
63. An English translation of the *Opera Inedita* by the present author, has been published by Macmillan at the same time as this volume. The introduction to the new edition contains biographical material on both Mayer and Lichtenberg, and a commentary on each of the treatises. The Latin titles of the four already mentioned are, respectively: "Observationes astronomicae quadrante murali habitae in observatorio Gottingensi"; "Methodus facilis et accurata computandi eclipses solares in dato loco conspicuas"; "De novo fixarum zodiacalium catalogo commentatio" and its appendage "Fixarum zodiacalium catalogus novus ex observationibus Gottingensibus ad initium anni 1756 constructus"; "De motu fixarum proprio commentatio."
64. "De variationibus thermometri accuratius definiendis", *ibid.*, pp. 3–10.
65. "De affinitate colorum commentatio", *ibid.*, pp. 33–42.
66. Several letters between Müller and Euler relating to this invitation are contained among "Der Briefwechsel L. Eulers mit G. F. Müller, 1735–1767", in *Die Berliner und die Petersburger Akademie der Wissenschaften im Briefwechsel Leonhard Eulers* (Berlin, 1959).
67. The original letter, dated Berlin 27 August 1754, is preserved in the Göttingen University library under Cod. MS. philos. 159 "Briefe von und an J. Tobias Mayer" (p. 7).
68. The original letter, dated St Petersburg 30 July 1754, is contained among the section of Mayer's correspondence that was transferred from the Göttingen Observatory to the local library in 1965 (*loc. cit.*, 5, Mayer 15_2).
69. A copy of this letter, dated 2 September 1754 is preserved in the University archive, Göttingen (*loc. cit.*, 26, p. 16).
70. Mayer's interest in the theory of machines, though aroused by his reading of Euler's "De usu machinarum maxime lucroso", *Commentarii Academiae Scientiarum Imperialis Petropolitanae*, x (1747), 67–94 [7], may well have been stimulated by his close contact with Segner. Equally significant, however, is the fact that he gave lectures on this subject in August and September 1752 as part of a course of public lectures on mechanics (*loc. cit.*, 5, Mayer 15_{34}). His experiments on the motion of a water-wheel and on the bending of beams, discussed in letters 7, 8, 9, are a direct outcome

of this aspect of his work to which no allusion has been made in this introduction since it has no positive bearing, either directly or indirectly, upon his researches in lunar theory.

71. A report on this lecture appears in the *Göttingische Anzeigen* . . . for 1759, pp. 1073–6 (cf. *loc. cit.*, 31).

72. That this *was* Mayer's intention is evident from remarks in a letter from J. D. Michaelis to W. P. Best, Göttingen 26 September 1754, (*op. cit.*, 74, p. 93).

73. J. Hadley, "The Description of a new Instrument for taking Angles", *Phil. Trans. R. Soc.*, xxxvii (1731) 147–57.

74. Four encouraging letters were sent by W. P. Best in London to Michaelis in Göttingen on 13 September, 8 October, 5 and 19 November 1754 whose contents the latter undoubtedly disclosed to Mayer. They are preserved under *Cod. MS. Michael. 320,* in the Göttingen University library (cf. pp. 555–69).

75. The bulk of the correspondence (over 50 letters), relating to Mayer's bid for a longitude prize, initiated on his behalf by Michaelis on 4 August 1754, is preserved in the *Staatsarchiv Hannover,* Hannover Des. 92 xxxiv No. II, 4, a^1. A further 35 letters pertaining to this theme and covering the same period 1754–65, are to be found among Michaelis's correspondence in the Göttingen University library (*Cod. MS. Michael. 320,* pp. 535–671). See also ref. 61.

76. Further information concerning the activities of this body can be found in E. G. Forbes, "The Board of Longitude Papers at the Royal Greenwich Observatory", *Actes du* XIe *Congrès International D'Histoire des Sciences,* iii (1968), 103–7.

77. Best to Michaelis, London, 11 February 1755, *loc. cit.*, 74, pp. 572–4. This letter has been published as no.73 in J. G. Buhle, *Literarischer Briefwechsel von J.D. Michaelis,* i (Göttingen, 1794). A copy is also preserved in *loc. cit.*, 75, pp. 67–9.

78. James Bradley to the Admiralty Secretary (Clevland), Greenwich, 14 April 1760 remarks: "The observations . . . were taken with the circular instrument constructed upon the plan proposed by Mr Mayer, though they were not made in the manner he intended, Capt. Campbell thinking it too inconvenient to attempt to do it on ship-board; but as the principal use of this construction is to obviate the inconvenience proceeding from the inaccurate division of instruments, and as that might be sufficiently removed by the care and exactness with which Mr Bird is known to execute those that he undertakes to make; a sextant of a radius twice as long as that of the circular instrument was made by him, and afterwards used by Capt. Campbell . . ." (cf. N. Maskelyne, *op. cit.*, 92, pp. cxiii–cxiv).

79. J. Ramsden, *Description of an Engine for Dividing Mathematical Instruments* (London, 1777); and *Description of an Engine for Dividing Strait Lines on Mathematical Instruments* (London, 1779).

80. W. E. May, "Early Nautical Instruments", *Naut. Mag.*, No. 145 (1945), 21–6. The advantages and disadvantages of Borda's design are discussed by Benzenberg (*op. cit.*, 1).

81. "Betr. den Contract wegen eines in London zu verfertigenden Quadranten zum Gebrauch des Observatorii astronomici 1754–55", *Personalakte Tobias Mayer* $4/V\frac{f}{2}$: (*loc. cit.*, 26).

82. The majority of these observations were published by Lichtenberg (*op. cit.*, 63, pp. 49–74). An accurate reduction of this catalogue to ecliptic co-ordinates was made by J. A. Koch, "Mayersches Zodiacal-Sternverzeichniss, auf den Anfang des Jahres 1800 reducirt . . .", *Astronomisches Jahrbuch für das Jahr 1790* (Berlin, 1787), pp. 113–44. The first complete catalogue, published by order of the commissioners of longitude, was *Astronomical observations, made at Göttingen, from 1756 to 1761* (London, 1826): its contents are critically discussed by Francis Baily in *Memoirs of the Astronomical Society of London*, iv (1831), 391–445. A revised edition appeared considerably later, entitled: *Tobias Mayer's Sternverzeichniß, nach den Beobachtungen auf der Göttingen Sternwarte in den Jahren 1756 bis 1760, neu bearbeitet von Arthur Auwers* (Leipzig, 1894).
83. Mayer states this explicitly in the introduction to his star catalogue (cf. Lichtenberg, *op. cit.*, 63, p. 46).
84. H. Quill, *John Harrison—The Man who found Longitude* (New York, 1966).
85. Details of this are given in *ibid.*, and in E. G. Forbes, "The Origin and Development of the Marine Chronometer", *loc. cit.*, 61, 1–25.
86. C. Niebuhr, "Über Längen-Beobachtungen im Orient u.s.w.", von Zach, *loc. cit.*, 1, iv (1801), 240–53.
87. Between 1752 and 1761, Mayer's wife gave birth to eight children (five sons and three daughters), of whom only the two eldest sons Johann Tobias (b. 5 May 1752) and George Moritz (b. 20 March, 1755) survived childhood.
88. Under the Act 5 George III, cap. xx.
89. This phrase was used by Maskelyne to distinguish the tables transmitted by Mayer's widow in the summer of 1763 from those sent by Mayer himself in 1755. It was, of course, the latter (earlier) set of manuscript tables that Maskelyne had used during his voyage to St Helena in 1761 to observe the transit of Venus, on which the lunar distance calculations published in his book *The British Mariner's Guide* (London, 1763) are based.
90. E. G. Forbes, "The Foundation and Early Development of the Nautical Almanac", *J. Inst. Navig.*, xviii (1965), 391–401.
91. The full title of this publication is "Tables requisite to be used with the Nautical Ephemeris for finding the Latitude and Longitude at Sea". An indication of the popularity and widespread use of the method of lunars may be gained from the fact that the 10,000 copies of this publication were exhausted within the next fourteen years, which led to a second edition being printed in 1781.
92. The equivalent Latin title is: "Tabulae motuum solis et lunae novae et correctae; auctore Tobia Mayer: quibus accedit methodus longitudinum promota, eodem auctore."
93. L. Euler, "Nouvelles recherches sur le vrai mouvement de la lune . . .", *loc. cit.*, 45, ix (1777), 1–94.
94. J. L. de Lagrange, "Essai d'une nouvelle méthode pour resoudre le problème des trois corps", *ibid.*, 1–126.
95. J. L. de Lagrange, "Sur l'équation seculaire de la lune", and P. S. de Laplace, "Sur le principe de la gravitation universelle, sur les inégalités séculaires des planètes qui en dépendent," are contained on pp. 1–61 and pp. 163–232 respectively of *Mémoires de Mathématique et de Physique. Année*

1773 (Paris, 1776). Laplace explicitly rejects Euler's hypothesis on p. 196 of the latter article.

96. *Mayer's Lunar Tables, improved by Mr Charles Mason* (London, 1780), p. 48.
97. P. S. de Laplace, "Sur l'Équation Séculaire de la Lune", *loc. cit.*, 39, *Année MDCCLXXVI* (1788), 235–64.
98. J. L. de Lagrange, "Recherches sur plusieurs points d'analyse relatifs à différens endroits des Mémoires précedens. Quatrième Mémoire. Sur l'équation séculaire de la Lune", *loc. cit.*, 17, MDCCXCII et MDCCXCIII (1798), 289–99.
99. D. Poisson, "Mémoire sur le mouvement de la lune autour de la terre", *Mémoires de L'Académie Royale des Sciences De L'Institut de France*, xiii (1835), 209–335.
100. See H. M. Nautical Almanac Office's publication entitled "A Modern View of Lunar Distances,", *loc. cit.*, 90, xix (1966), 131–53.

CORRESPONDENCE

1. *Mayer to Euler; Göttingen, 4 July 1751*

Dear Sir,
Honourable Professor!

The desire to become still better acquainted with you, sir, than through your uncommonly fundamental treatises, which is as old as the happy occasion in which these treasured writings first came into my hands, could finally not be withheld any longer although I have made a big effort not to [write to you]. Both the flattering idea of the advantage which I could promise myself out of this liberty which I have taken, as well as the kindness which you, sir, have shown in answering the letters of my brother-in-law Professor Lowiz in Nuremberg,[1] has largely suppressed the fear that you, sir, could interpret my presumption unkindly and gave me hope that my letters may perhaps meet with the same fortune.

Each time that I read through the theory of light and colours which you, sir, have given in your *Opuscula*,[2] I appreciate more and more the advantage of these theories as compared with Newton's. My researches concerning the refraction which objects on our earth suffer in the atmosphere, have led me to a fact which can perhaps be useful for explaining a difficulty that turns up in this theory without really belonging to it. You, sir, determine with Newton the speed of sound from the altitude of the column of air of 26,000 Paris shoes,[3] so that this speed is equal to that of a body which falls from half of the estimated height and thus incidentally 900 Paris shoes in 1″. However, this calculated velocity is roughly about 140 shoes smaller than experiment gives. What I now have to state in this connection is the following. If y represents the elasticity of the air at every altitude x above the Earth's surface, $le = 1$,[4] however, then $y = e^{-x/26,000}$, if x is expressed in Paris shoes. The experiments which have been made with the barometer on various mountains agree perfectly with this equation.[5] According to the law that the elasticity of the atmosphere is proportional to its density, the previous equation should also apply if y denotes the density of the atmosphere at the altitude x. Yet I have found through experiments which have been made on the refraction of objects on the Earth,[6] as well as from astronomical refraction, that in the latter case, namely where y denotes the density of the air, one should set $y = e^{-x/36,000}$, or if x is taken to be in Rhein shoes, $y = e^{-x/37,260}$.[7] The speed which a body falling from half of the coefficient of x, namely from 18,630 shoes, attains is now 1079 Rhein or 1042 Paris shoes in 1″, and consequently exactly equal to the velocity of sound. Now as before, the decrease of the elasticity of the air depended upon the height of the natural column of air (26,000 shoes), whose weight is equal to the elasticity itself; the decrease of the density, however, depended upon a greater height (36,000). Thus it can perhaps be a fact of Nature that the speed of sound must be determined not from the first [law], but from the last. [Since I would like to know] whether this supposition may be substantiated, or whether I should put it down to a pure chance, that is why, dear sir, I have taken the liberty of transmitting this fact to you.

Your tracts on analysis and mechanics, honourable sir, have enabled me to study the Moon's motion according to Newton's theory,[8] something which I tried vainly to do several times beforehand. For this purpose I used a method of approximation which is largely in accordance with that which you, sir, have used for the investigation concerning the motion of Saturn and Jupiter,[9] except that I have always retained the mean anomalies of the ☽ and ☉ instead of the eccentric anomalies. In this way the calculation of the Moon's orbit from the tables becomes slightly simpler, since the arguments should only be roughly determined from the mean motions, so that I do not once require the actual position of the Sun for the calculations of the Moon's position. While I am still unable to say that my observational result is completely satisfactory, I certainly do not believe on this account that Newton's law of attraction should be amended, as Mr Clairaut[10] believes; there being so many circumstances which make the calculation and approximation doubtful, that the error can rather be on the latter side. Besides, I have noted that it is extremely difficult, or perhaps quite impossible, to establish the truth or falseness of Newton's law of attraction from the motion of the Moon. After I had therefore accepted as true [the fact] that the Moon's orbit cannot be accurately enough determined by the previous method, it occurred to me that one might predict the Moon's position more accurately by a simple method [applicable at] any time using the periods of 223 lunations which Halley has proposed for calculating the eclipses of the ☽ and ☉.[11] I have attempted this by means of my previously calculated equations of the ☽'s mean position and obtained the following: if at a given time T the Moon's true position in its orbit L is known from observations, and for just this time from the tables

the mean anomaly of the ☽ $= p$
the mean anomaly of the Sun $= s$
the mean distance of the Moon from the Sun $= \omega$

the true position of the Moon in its orbit for the time $T + 18^{y}\ 11^{d}\ 7^{h}\ 43'\ 30''$ or $T + 223$ synodic months $=$

$$+ L + 10^\circ 48' 16'' + 1133'' \cos p + 28'' \sin p - 77 \cos 2p$$
$$- 4 \sin 2p + 5 \cos 3p \qquad \text{I}$$
$$+ 124 \cos s - 11 \sin s - 4 \cos 2s \qquad \text{II}$$
$$- 26 \cos (p - s) - 3 \sin (p - s) \qquad \text{III}$$
$$- 14 \cos (p + s) + 1 \sin (p + s) \qquad \text{IV}$$
$$- 226 \cos (2\omega - p) + 6 \sin (2\omega - p) + 3 \cos (4\omega - 2p) \qquad \text{V}$$
$$- 9 \cos (2\omega - 3p) \qquad \text{VI}$$

$- 2 \cos (\omega - p) + 20 \cos (2\omega - 2p)$ VII
$- 13 \cos (2\omega - p + s) + 1 \sin (2\omega - p + s)$ VIII
$- 36 \cos (2\omega - p - s) - 3 \sin (2\omega - p - s)$ IX
$+ 25 \cos (2\omega - 2p + s) - 4 \sin (2\omega - 2p + s)$ X

I have introduced these 10 equations into very many tables and compared them with the observations, which have always been accurate to within 10″ or 20″. Consequently, if one had a continuous series of observed positions of the Moon throughout 18 years, one could determine the Moon's position very accurately for any future time by means of these tables. Lastly, when solving a certain problem, I have come across an equation which, though it would appear to be simple, I am unable to integrate. It is namely the following: $dy = \frac{dx}{\log x}$, or this: $dz = e^{e^x}\, dx$, where $\log e = 1$. In the treatise in which you, sir, have explained Mr Robins' rules of artillery[12] there is a similar equation on page 131 which, however, is only integrated by approximation, an advantage not permitted to me with the above equation. I could not let pass the opportunity of seeking your advice regarding this equation, sir, and I beg you not to interpret such [a desire] unkindly.

It is perhaps not unknown to you, sir, that we here in Göttingen are getting an *Observatorium Astronomicum*. The installation itself has been jointly commissioned to Professor Segner and myself, and the building has now been actually begun, as the instruments belonging to it are also already under construction.[13] As soon as a few of the latter are ready, I shall begin to make corresponding observations on the parallaxes of the Moon and Sun with the Abbé de la Caille, who has gone off to the Cape of Good Hope.[14] Concerning the former [i.e. the parallaxes of the Moon], I already made a few observations of my own while I was still in Nuremberg, through which I soon found the parallaxes of the Moon very accurately, and I would have perhaps found them even more accurately had this very investigation not led me to a circumstance which seems to make the matter doubtful, and lies in the following:

As I have used two different methods to determine the actual amount of the Moon's parallax, or rather of its ratio to the diameter itself, the amount of the parallax found from both methods ought to have been the same. Yet this did not occur, but the first method gave me the ratio of the parallax to the diameter as 55 to 30; against that, the other showed this same ratio as $55\frac{2}{5}$ to 30. In both methods I took account of the true figure of the Earth, thus the discrepancy cannot enter on this account.[15] Meanwhile I discovered that, if I supposed that the centre of gravity of the Earth, around which the Moon moves, is distinguished from the centre of magnitude or figure [i.e. the geometrical centre]—the former falling within the noon hemisphere of the Earth—then I say the results of both methods would come out to be the same. I am, however, far too timid to dare to state with any certainty whether this difference of the centres of the Earth conflicts with the laws of Nature, without having made further researches and a great deal more observations too. So much is certain, that, if this irregularity does occur, it could be regarded as one of the major reasons why the Moon's motion could not hitherto be determined exactly from theory, and it therefore seems to be worth the trouble to consider still further experiments. The observations which the Abbé de la Caille has resolved upon making, are not likely to enable one to decide on this point.

Now my letter has grown to be so much over the usual length, that I am almost afraid of presenting it to you, sir. If you find nothing in it worthy of your attention, you will certainly see from its length, sir, how great an effort I have made to be a worthy correspondent. Should I be lucky enough to receive a favourable reply, I shall consider my effort more rewarded than it deserves. I remain with the greatest possible esteem,

Sir,

Your obedient servant,

T. Mayer

2. *Euler to Mayer; Berlin, 27 July 1751*

Your kind letter has awakened in me a pleasure which is particularly great since I have already seen some time ago such magnificent proofs of your deep understanding in the most important and most difficult questions of mathematics and more especially astronomy.[1] The splendid intention of the Göttingen Academy to build an observatory could not be better put into effect than through your skill in this science, from which for good reasons one can rightly expect the most wonderful discoveries. I therefore treasure the honour of your acquaintance very highly, and for my part I shall, with the greatest pleasure, do all that I possibly can to retain it.

Your admirable remarks concerning the unequal relation according to which the density and elasticity of the air above the Earth decreases, are indeed of the highest importance to natural science and deserve to be thoroughly investigated. Especially, the cause of this inequality must lead us to an exact knowledge of the nature of the atmosphere which will enable us to better establish many of its properties. Indeed, the propagation of sound, of which I must confess that my determination is still very uncertain, is significant in this connection. If you already have the 1st volume of the *Novi Comment. Academiae Petrop.*, you will find in it a treatise by me "De propagatione pulsuum per medium elasticum",[2] in which I have discussed the difficulty of this question much more than its solution. Meanwhile, it seems to me that the speed of sound depends merely upon the density and elasticity of the air at the place through which the sound passes, and I cannot yet foresee what the variation in the altitude could contribute to it. Also, the exact agreement of your formula with experiment seems to give no little weight to the matter: however, I believe that in connection with this question one must make a distinction between the propagation of a single pulse and a series of many successive pulses: and I am almost willing to believe that, if it were possible to excite one single pulse without others following it immediately, this [i.e. the pulse's speed of propagation] would not amount to much more than 900 shoes in a second. If, however, many pulses succeed one another, it seems very probable to me that their speed will consequently be increased. I have explained this in more detail in a continuation of my *Opuscula*,[3] which has been published here, and although I am unable to explain with clarity the increase of speed *a priori*, it nevertheless seems to me as if the air's [own] speed would account for the discrepancy. After having given closer consideration to the different refraction of the various rays of light, I am unable to find any other reason than that the pulses travel more quickly in one ray than in the other: but now I find no other significant differences in the beams apart from the frequency of the pulses, therefore I conclude that the speed of the pulse depends not only on the density and elasticity of the media, but also on the frequency of the pulses. However, I put forward these views only as a rough conjecture, which nevertheless seems to deserve further consideration.

I note with particular pleasure that you have made such good progress on the lunar theory. After considering all the difficulties you have certainly taken the best means of determining the difference between the present position of the Moon and that where it has been 223 lunations beforehand; and since you noticed in this way that the error which always persists in the observations does not exceed 20″, you have certainly achieved in this all that can ever be desired. For, since one can always know for the past 18 years how much the true position of the Moon differs from the tables, one can accurately determine from these very tables (bad though they may be, in themselves) the position where the Moon stood in the sky 18 years 11 days 7 hours 43′ 30″[4] before, and consequently correctly determine the present position of the Moon. I have also indicated that for the particular purpose of determining the inequalities of the motion of the Moon it would be easier to calculate the mean longitudes and mean anomalies of the Sun and Moon; however, since the true distance of the Sun and Moon is an important factor in the calculation, the formulae for the sine and cosine of the true distance become so complex, if the mean is introduced instead of the same [i.e. the true distance], that in my opinion you do not gain much thereby. And I also believe that part of your equations would be shorter, and would be fewer in number, if p denoted the true anomaly of the Moon, and ω the true distance of the Moon from the Sun. At least, when I repeated my researches on this subject, I took into account the true anomalies of the Sun and Moon as well as the true distance of the Moon from the Sun, which seems to me to be far more convenient. My actual intention was to investigate thoroughly how accurately the theory of Newton concerning the motion of the Moon's apogee agrees with the observations. For not only I, but others too who worked on this, have hitherto always found that according to theory the motion of the apogee should be only half as large as it has, in fact, been found to be. Now, however, I have found to my great pleasure that this part of Newton's theory is exactly in accordance with the observations: therefore it is much less to be doubted that the agreement should not be complete: there are, however, still nearly insurmountable difficulties connected with the calculation for determining accurately all the lunar inequalities from the theory.

This equation $\mathrm{d}y = \frac{\mathrm{d}x}{\mathrm{l}x}$ is in no way integrable, and if one does not want to use the approximation I know of no other method than that the value of y be determined through the *Quadrature Mechanice* [viz., an empirical, or graphical, method of integration] which can in fact be regarded simply as an approximation, and a very incomplete one at that. I wanted to determine the value of y from the following set-up through a series;

$$y = \frac{x}{\mathrm{l}x} - \int x\,\mathrm{d}\frac{1}{\mathrm{l}x} = \frac{x}{\mathrm{l}x} + \int \frac{\mathrm{d}x}{(\mathrm{l}x)^2};$$

further

$$\int \frac{\mathrm{d}x}{(\mathrm{l}x)^2} = \frac{x}{(\mathrm{l}x)^2} - \int x\,\mathrm{d}\frac{1}{(\mathrm{l}x)^2} = \frac{x}{(\mathrm{l}x)^2} + 2\int \frac{\mathrm{d}x}{(\mathrm{l}x)^3}.$$

Further

$$\int \frac{\mathrm{d}x}{(\mathrm{l}x)^3} = \frac{x}{(\mathrm{l}x)^3} - \int x\,\mathrm{d}\frac{1}{(\mathrm{l}x)^3} = \frac{x}{(\mathrm{l}x)^3} + 3\int \frac{\mathrm{d}x}{(\mathrm{l}x)^4}$$

and so on, from which one finally obtains

$$y = \frac{x}{\mathrm{l}x} + \frac{x}{(\mathrm{l}x)^2} + \frac{1.2x}{(\mathrm{l}x)^3} + \frac{1.2.3.x}{(\mathrm{l}x)^4} + \frac{1.2.3.4.x}{(\mathrm{l}x)^5} + \text{etc.}$$

Then, if $x = 0, y = 0$: but if $x = 1, y = \infty$, from which it follows that since x cannot be negative (otherwise $\mathrm{l}x$ would be imaginary), there cannot be other values to seek for y except those for which x is less than unity; and then lx becomes a negative quantity. Thus $\frac{1}{\mathrm{l}x} = -U$; and so[5]

$$y = -Ux(1 - 1.U + 1.2U^2 - 1.2.3.U^3 + 1.2.3.4U^4 - \text{etc.})$$

and then the whole problem depends upon the sum of the bracketed series which, since it is strongly divergent, is subject to new difficulties. I have, however, discovered that the series

$$1 - 1U + 2U^2 - 6U^3 + 24U^4 - 120U^5 + 720U^6 - \text{etc.} = S$$

is equal to the following continuous fraction

$$S = \cfrac{1}{1 + \cfrac{U}{1 + \cfrac{U}{1 + \cfrac{2U}{1 + \cfrac{2U}{1 + \cfrac{3U}{1 + \cfrac{3U}{1 + \cfrac{4U}{1 + \cfrac{4U}{1 + \cfrac{5U}{1 + \cfrac{5U}{1 + \text{etc.}}}}}}}}}}}$$

which always closely determines the limits of the value of S, and thus one can approximate to the value of S as closely as one will. Then the values approximating to S are:

$$1;\ \frac{1}{1+U};\ \frac{1+U}{1+2U};\ \frac{1+3U}{1+4U+2UU};$$

$$\frac{1+5U+2UU}{1+6U+6UU};\ \frac{1+8U+11UU}{1+9U+18UU+6U^3};$$

$$\frac{1+11U+26U+26UU+6U^3}{1+12U+36UU+24U^3};\ \frac{1+15U+58UU+50U^3}{1+16U+72UU+96U^3+24U^4};\ \text{etc.},$$

of which every alternate one is greater than S. Thus

$$S < \frac{1}{1};\quad S > \frac{1}{1+U};\quad S < \frac{1+U}{1+2U};$$

$$S > \frac{1+3U}{1+4U+2UU};\quad S < \frac{1+5U+2UU}{1+6U+6UU};$$

$$S > \frac{1+8U+11UU}{1+9U+18UU+6U^3}$$

and so forth.

For the sake of brevity, one sets the sums: $1 + U = A$; $1 + 2U = B$; $1 + 4U + 2UU = C$; $1 + 6U + 6UU = D$; etc. So that $A = 1 + U$; $B = A + 1U$; $C = B + 2UA$; $D = C + 2UB$; $E = D + 3UC$; $F = E + 3UD$, etc. Whence it is evident how the values of the letters A, B, C, D, E, etc. may be found in terms of the previous expressions: having found these, one has the convergent series:

$$S = 1 - \frac{1U}{1A} + \frac{1.1U^2}{AB} - \frac{1.1.2U^3}{BC} + \frac{1.1.2.2.U^4}{CD} - \frac{1.1.2.2.3U^5}{DE} + \frac{1.1.2.2.3.3.U^6}{EF} - \text{etc.}$$

or when two terms are always taken together:

$$S = \frac{1}{A} + \frac{1.U^2}{AC} + \frac{1.4U^4}{CE} + \frac{1.4.9.U^6}{EG} + \frac{1.4.9.16.U^8}{GF} + \text{etc.}$$

or also

$$S = \frac{1}{2} + \frac{1}{4A} + \frac{1 + U - UU}{8AB} + \frac{1 + 4U + UU - 6U^3 - 4U^4}{16ABC} + \text{etc.}$$

This series is, however, not convergent. Having found S in this way, since $U = -\frac{1}{lx}$, $y = -UxS$. Thus for the case $x = e^{-1} = \frac{1}{e}$; and $U = 1$, $S = 0{\cdot}5963473621237$ and $y = \frac{-5}{e} = \frac{-292}{1331}$ nearly. In such a manner, for a particular value of x, the corresponding value of y can be found with as high an accuracy as is desired, without too great a difficulty.

Your discovery that different methods do not give the same ratio between the parallaxes of the Moon and its diameter, is without the slightest doubt of the greatest importance in astronomy; and these [methods] can render no greater service to this science as when they allow the reason for the discrepancy to be established through very careful observation. The same [reason,] from which should follow the kind of parallactic angle that will be deduced from it, must undoubtedly lie hidden in the nature of the two methods; of the two lines which form the angle, one certainly goes from the centre of the Moon into the eye of the spectator. However, where the other goes, whether towards the centre of the Earth, or towards the point where the vertical line from the position of the observer intersects the Earth's axis, or towards another point, the method must show. Besides this, the question also arises as to whether according to both methods the angular parallax is really formed at the centre of the Moon, or in its surface, since this could also give rise to a discrepancy. What the true centre of gravity of the Earth could contribute towards, I cannot really comprehend: it would then be the case that the distance of the Moon from the Earth was deduced from theory; for the theory, but not the observations, relates to the centre of gravity. However, there is no one better able to resolve this problem than yourself.

Please give my best regards to Professor Segner, from whom I have not received a written communication for a long time, and thus I do not know whether perhaps my letter to him went missing.

I have the honour to remain with the greatest respect

Your obedient servant,

L. Euler

3. *Mayer to Euler; Göttingen, 15 November 1751*

Dear Sir,
Honourable Professor,

You have done me a great favour through your very kind answer to my letter, not only because I have found in it the solution to my previous doubts, but also because I have thereby been granted leave to continue this correspondence which is so advantageous to me. Although I very much doubt already whether this will always contain something which would be worthy of your attention, I could not but make use of the privilege which I have once received from you.

You will have noted, sir, from my previous letter, that I have been struggling with the refraction of our atmosphere. From this, however, I cannot clearly see how the propagation of sound ought to depend upon the decrease of the density of the air at higher altitudes. Since that time, I have given more careful thought to this matter, [and] I have found that the previously-mentioned formula regarding the densities can still vary considerably from the truth, and that the densities perhaps decrease in the same ratio as the elasticities of the air, or as the height of the mercury in the barometer. I have determined the formula in question, or rather the coefficient 6000 which it contains, from another formula which I had previously calculated for the astronomical refraction; since, however, this is not complete, although it already agrees quite well with the observation, neither do I trust that any longer, and thus I am abandoning my previous considerations. I have encountered so many difficulties in the discovery of my formula for the astronomical refraction and still more for the refraction on the Earth that, regardless of all my efforts, I could still not find anything suitable, and thus I wished so much that you, sir, at your convenience, would take over these matters, part of which you have already explained in the Berlin Calendar of 1751.[1]

I have read through your ideas on the propagation of sound and light in the extension of your *Opuscula*[2] with the greatest pleasure, and the matter appears so natural to me, that I certainly attach more than a rough probability to it. In the third part of this work I have also found the incomparable theory of the magnet.[3] I doubt whether one could wish for more in the explanation of the properties of this stone than you have supplied, and I now no longer doubt that all other difficult questions in the study of nature ought ultimately to be capable of being solved in a natural manner. In §24 of this theory I have come across a small difficulty which, sir, I am unable to conceal. If universal gravitation were produced by a magnetic pole, in my judgment the attraction would not remain in the direct ratio of the masses, as one normally assumes, but would be related to the large number of magnetic particles in a heavenly body. Now, since it is unbelievable that the quantity of these particles is proportional to the whole mass of a heavenly body, the direct ratio of the masses would no longer apply, and if none of this material existed in the neighbourhood of two heavenly bodies, there would therefore be no attraction between them, etc. I once read long ago an article by someone called Müller,[4] in which the author explains the cause of gravity as due to the mere pressure of the aether, and if one grants him his principle, this gravity follows naturally from it, although he certainly presents his thoughts in a very confusing manner. Indeed, I cannot say that he has particularly impressed me, yet I think that his paper still deserves some attention. Perhaps you have seen this book, in which case I should very much like to ask you your opinion of it.

My present academic duties now no longer leave me enough spare time to do much regarding the lunar theory. However, I have not lost the desire to do so. I have noticed with great pleasure that you have fortunately verified the Newtonian law of attraction through the determination of the Moon's apogee, and astronomers have taken up the position of adopting this law with more certainty in other investigations. How troublesome would even the solution of the most trivial question have been if, instead of this very simple law, it had been necessary to have used another undoubtedly more complicated one? In my calculation I could not determine the motion of the apogee quite accurately enough; it seemed in the same calculation as if this motion depended not only upon the action of the Sun, but also simultaneously very markedly upon the eccentricity of the Moon's orbit, and therefore I became doubtful whether my calculation had the necessary accuracy to decide this question. For this reason, I then had a glance at the equations of longitude and of the distance of the ☽ from ♁. I have now actually not quite succeeded in finding the right solution in this way, in that I could not completely determine some equations from the theory, particularly those which depend upon such angles as $\omega - p$, $2\omega - p$, $2\omega - 2p$, $3\omega - 2p$, $3\omega - 3p$, etc., where ω signifies the Moon–Sun distance and p the mean anomaly of the Moon. However, I nevertheless had so much success thereby, that if I had to repeat the calculation again, I would have the hope of obtaining everything much more correctly and precisely, with less trouble. But, all my work would be in vain if you, sir, should wish to publicise your discoveries in this matter.

As I read in the newspapers, M. Clairaut has obtained the prize of the Petersb[urg] Academy relating to this problem,[5] and at the same time I understand that he discovered that the motion of the apogee also conforms with the theory. If the Academy wanted to let others have his paper by printing it, it [i.e. the Academy] would bring about much good; yet you yourself, sir, could accomplish these same commitments at a more advanced level, for your investigations in this subject certainly exceed everything which others were able to do in it.

It is not unknown to you, sir, that the nutation of the Earth's axis is caused by the attraction of the Moon on our spheroidal Earth. Since this effect cannot occur without the Moon's also being rotated at the same time in its orbit, it would be very good if someone would calculate the inequalities which originate in the Moon's motion due to the shape of the Earth. As these can only be very small, I believe that the calculation would not be quite so difficult. Perhaps you have done this already.

Mr Mylius has recently made several objections in his *Physikalischen Belustigungen* against my proof that the Moon has no atmosphere,[6] and quoted in this connection your observation which is in the Memoires of the Royal Academy of that place [Berlin][7] to his advantage. I have not yet made up my mind whether to defend myself, although it would be very easy for me. I shall not do so if I know that you may not like it. Although I could also continue this controversy without offending anyone in the world, time and leisure are much too valuable to me for me to spend them on a matter whose importance I am unable to assess.

I have the honour to remain with all imaginable respect

Your obedient servant,

T. Mayer

4. *Euler to Mayer; Berlin, 25 December 1751*

Right Honourable Sir,
Most Highly Esteemed Professor,

I have already tried long ago at different times to determine clearly the astronomical refractions in our atmosphere solely from the theory of the aether. I have, however, come across so many difficulties in the equations thus deduced, that I have almost given up all hope of proceeding further with this research; meanwhile, you will have seen M. Bouguer's treatise on this matter,[1] in which he already deduces the table of refraction, although the theoretical investigation does not appear to me to be sufficiently complete. The whole difficulty does not actually lie so much in the theory itself, as in the analysis of the differential equations which cannot be integrated in such a manner that the approximation will be reliable for all angles of obliquity.

As for the cause of gravity which I have rejected, I consider the objection that the attraction must not necessarily be proportional to the masses, to be of no great importance, as it is still not decided by any single phenomenon that the attractive forces of heavenly bodies are proportional to their masses. On the contrary, Newton tried to determine the masses on this basis since there is no other way of specifying them. As soon as one now places the statement that the attractive forces are proportional to the masses (which is founded on a crude hypothesis) in doubt, this objection against my idea is completely eliminated. Nevertheless, I willingly admit that this explanation of mine is still too unreliable, and content myself that I have thereby demonstrated that gravity could have a mechanical cause, although I greatly fancy that the impossibility of this has now been sufficiently established. I have indeed read the treatise of Mr Müller concerning this matter,[2] but all that I am able to remember about it is that his proposed explanation is in direct conflict with the mechanical principles, since it is based upon [the assumption] that the aether existing between two heavenly bodies has a weaker elastic force, which is quite opposed to the principles of equilibrium. On the other hand, it is certain from mechanical principles that where an elastic fluid is in motion, its elastic force will be diminished. Hence it can be shown that if the aether surrounding heavenly bodies were to be set in motion, it is conceivable that gravitation could originate in this; however, as long as the aether is at rest, this is impossible.

The theoretical determination of the motion of the lunar apogee is one of the most difficult parts of this theory, since it depends upon all the inequalities of motion, hence this must have been known previously, before the motion of the apogee can be determined. In particular the true amount itself originates from the inequality dependent upon the angle $= \omega - p$, according to your notation. I have sent my treatise upon this to Petersburg, where it will be printed with Mr Clairaut's prize essay.[3] According to the theory of Newtonian attraction, the centripetal force of the Earth, owing to its spheroidal shape, ought certainly to decrease in about the inverse ratio of the distance squared, and when z denotes the distance, the formula for this force must therefore be $\frac{A}{zz} + \frac{B}{z^4}$ when the Moon is in the equator. If, however, it has a declination, this would also still enter into this formula. Hence I am almost sure that the lunar inequalities that still remain, which could not be determined through any theory, arise from this reason. Indeed, the term $\frac{B}{z^4}$ would affect the coefficients of the inequalities only rather slightly, without yielding new [ones]. Only when the Moon is not in the equator, must an inequality be available which depends on that declination. When the shape of the Moon itself is also non-spherical, special inequalities arise in that motion on this account, which, however, principally affect only the motion of the apogee.

If you, in regard to the lunar atmosphere, do not simply deny the phenomenon that at annular eclipses of the Sun the solar disc during the annular phase seems significantly larger than beforehand or afterwards, as I have quite clearly observed, I am completely satisfied. It then follows incontestably from that, that the solar rays passing so closely to the Moon were somewhat curved. Now I will not dispute whether this bending was produced by a lunar atmosphere, or, as some want, by the Moon's attraction, [just] as the followers of Newton account for the bending of light. Only, since I do not believe the attraction to be an internal quality of bodies, I remain of the opinion that the bending of light also arises from a particular state of the aether surrounding the body, to which I have no hesitation of referring under the name of an atmosphere, and in this sense I also maintain that an atmosphere may exist around the Moon which, as I have indeed indicated, is so subtle that it seems scarcely to deserve this name. I assume, therefore, no other part in this dispute, than insofar as the phenomenon of the annulus will be admitted or denied.

I have the honour to remain with all respect

Your obedient servant,

L. Euler

5. *Mayer to Euler; Göttingen, 6 January 1752*

Dear Sir,
Honourable Professor.

Since you have found the matter concerning the determination of the refraction of our air [to be] so difficult, I make no wonder that I was still less able to succeed in it. I have also always investigated this in all sorts of ways, yet the most worthwhile result that I got out of it was when I took the curve of density to be a straight line, or, if x is reckoned as the height above the Earth's surface, and y the density of the air at this altitude, $y = A - Bx$. Then, in this case, the formula for the refraction may be integrated by simple logarithms, or even algebraically by means of a suitable approximation. In order, however, to avoid absurdities, the integral should be taken no further than to $x = \frac{A}{B}$. Now, although this hypothesis is an open contradiction of the truth, the resulting formula for the refraction nevertheless agrees very closely with observations, and this makes me think that it would not really help much if one were to choose straightaway another and better expression for the density of the air. I have already seen Mr Bouguer's[1] treatise ten years ago in Stuttgart at H. Bilfinger's and where it was rightly [placed] among the Prize Essays of the Paris Academy, however I had not an opportunity of examining the whole but only to make a few excerpts from it. His formula can also be expressed: $\rho = C \arcsin (D \sin \alpha) - C\alpha$, where ρ denotes the refraction, but α the distance from the apparent zenith. By approximation, however, one obtains from this formula

$$\rho = C\sqrt{(\cot^2 \alpha + 2 + 2D)} - C \cot \alpha$$

My formula now gives nearly the same thing, and appears thus

$$\rho = m \tan \tfrac{1}{2}\alpha[1 - \beta \cos \alpha + \sqrt{(\beta^2 \cos^2 \alpha + 2\beta + 1)}]$$

and if one puts $\beta = n - 1$, it transforms into the following

$$\rho = m[\sqrt{(n^2 \cot^2 \alpha + 2n - 1)} - n \cot \alpha + \sin \alpha + n \cos \alpha \cot \alpha - \sqrt{([\sin \alpha + n \cos \alpha \cot \alpha]^2 - 1)}]$$

in which the first two terms are in complete agreement with Mr Bouguer's formula, but the last always remain very small. Which strengthens even more my above-mentioned view, since Mr Bouguer has undoubtedly adopted another hypothesis. Otherwise, $1:1 - m$ is the ratio of the angle of incidence to the angle of refraction of a ray coming from the vacuum into the lowest [layers of] air; but β will be determined from the above equation $y = A - Bx$ through the constant coefficients A and B, and on the basis of my researches it is equal to about 545, or $\frac{\textit{radius of the Earth in Paris feet}}{36{,}000}$, where 36,000 is the number of which I informed you in my first written communication. According to Mr Bernoulli[2] one must assume for it 22,000 which, however, is contrary to experience; and if the densities decrease in exactly the same ratio as the elasticities, the same divisor would be equal to 27,000. Hence all the variations in the astronomical refraction, which are dependent upon the different temperatures of the air, can now be simply derived.

I have never been able to persuade myself of the idea of attributing the attraction to something belonging to bodies, and the last objection against your explanation was itself not something stemming from such a ground. But because you have made the first approach to a rational explanation, I was desirous of seeing this very difficult material investigated more closely by yourself, if possible, and I now perceive that what Newton specified as the mass of the heavenly body rests merely on a hypothesis. It has occurred to me through the study of your treatise on the propagation of light[3] [to enquire] whether the agitation of the particles of aether arising from this propagation could not produce a decrease in its [i.e. the aether's] elasticity. It seems to me that this is very probable, although I do not trust myself to maintain anything definite about it, but submit these ideas to your very friendly criticism.

I have recently brought my researches on the Moon's parallax into order and shall give them over to our local Royal Society.[4] I find thereby the diameter of the moon at its mean distance $= 31'\ 10''$ and the horizontal parallax, or rather the semi-diameter of the Earth's equator observed from the centre of the Moon, $= 51'\ 10''$. I believe that the amount of the error in this latter [figure] never exceeds $10''$. I tried to determine this very parallax from the theory, and after I had taken account of most of the principal factors, it [i.e. the parallax] at the mean distance came out $= 57'\ 10''$; which agrees extremely well with the above-mentioned values found from observations. In this [calculation] I took the mass of the Moon $= \frac{1}{70}$ mass of the Earth, as supplied by Mr Bernoulli's determinations.[5] I have also been mindful of the Earth's spheroidal shape which I have taken into consideration at the same time, and put the attraction of the Earth on the Moon $= \frac{1 - 3n/5}{z^2} + \frac{3n}{5z^4}$, where the semi-diameter of the Earth's equator $= 1$, the half-axis $= 1 - n$, and the gravitation or attraction below the equator $= 1$, although the term $\frac{3n}{5z^4}\, k\langle 1\rangle$ has a marked influence on the distance of the Moon from the Earth, or on the parallax.

This last term is just the one which can produce some inequalities in the Moon's motion. I presume that, if one takes into consideration at the same time the Moon's declination, then not only one but more inequalities must arise, which simultaneously depend on the angles ω and p; and some of them can also be still more marked, provided that [the inequality] which is due only to the declination, is still appreciable. Also, I always suppose that there are still new equations of the angle $A\omega - \begin{pmatrix} A-1 \\ 0 \end{pmatrix} p \pm Bs$; then if, for example, such an angle were always constant or only very slightly variable, an infinitely long or at any rate very long equation of the Moon's motion would result. Perhaps the approximation could be so arranged that one would concentrate only on those angles alone; the others, however, which converge more rapidly, quickly disappear. By means of my solution I have found the following inequalities:

	For the Moon's orbital position mean longitude				*	For the Moon–Earth distance 100,000 + 148	
I	−	6°	18′ 56″	$\sin p$	(22736″)	+	$5454 \cos p$
	+		12 52	$\sin 2p$	772	−	$148 \cos 2p$
	−		0 32	$\sin 3p$	32	+	$6 \cos 3p$
II	+		11 20	$\sin s$	(+680″	−	$14 \cos s$
	−		0 10	$\sin 2s$	−10		
	−		1 46	$\sin \omega$	−106	+	$24 \cos \omega$
III	+		38 10	$\sin 2\omega$	+2290	−	$750 \cos 2\omega$
	+		0 17	$\sin 4\omega$	+17	−	$2 \cos 4\omega$
IV	+		1 54	$\sin (p - s)$	(+114	−	$27 \cos (p - s)$
V	−		1 42	$\sin (p + s)$	(−102	+	$25 \cos (p + s)$
	−	1	15 40	$\sin (2\omega - p)$	−4540	+	$1000 \cos (2\omega - p)$
VI	+		0 33	$\sin (4\omega - 2p)$	+33	−	$7 \cos (4\omega - 2p)$
VII	−		0 36	$\sin (4\omega - 3p)$	−36″	+	$6 \cos (4\omega - 3p)$
VIII	−		0 43	$\sin (4\omega - p)$	−43	+	$11 \cos (4\omega - p)$
IX	−		1 44	$\sin (2\omega + p)$	−104	+	$26 \cos (2\omega + p)$
X	−		0 37	$\sin (\omega - p)$	−37	−	$1 \cos (\omega - p)$
	+		3 20	$\sin (2\omega - 2p)$	+200		$\cos (2\omega - 2p)$
XI	−		1 0	$\sin (2\omega - 3p)$	−60	−	$50 \cos (2\omega - 3p)$
XII	+		0 23	$\sin (2\omega + s)$	+23	−	$8 \cos (2\omega + s)$
XIII	−		0 10	$\sin (2\omega - s)$	−10	+	$51 \cos (2\omega - s)$
XIV	−		0 55	$\sin (2\omega - p + s)$	−55	+	$12 \cos (2\omega - p + s)$
XV	+		4 31	$\sin (2\omega - p - s)$	+271	−	$55 \cos (2\omega - p - s)$
XVI	+		1 28	$\sin (2\omega - 2p + s)$	+88	+	$2 \cos (2\omega - 2p + s)$

* All the numbers in this column were introduced by Euler.

Among these many inequalities there are some which I can completely determine, and some which I must correct, through observations. I have completely discarded still more on account of their smallness. The angles ω, p and s invariably denote the mean motion, which in fact brings several advantages not in the solution, but in practice, because in such a manner the arguments of the inequalities can be calculated more simply. I have indeed always supposed, yet could never be certain, that the inequality associated with the angle $2\omega - p$, which is without doubt difficult to determine and which was explained by Newton as due to the variation of the eccentricity of the Moon's orbit,[6] strongly affects the motion of the apogee. Now, I want particularly to make new attempts to determine the motion of the apogee, and to see whether I do not arrive at it if instead of the above-mentioned inequality I take the eccentricity as being really variable. Meanwhile, I eagerly await the treatises of yourself and Mr Clairaut.[7] You would greatly oblige me if I could obtain through your assistance a copy as soon as they are published, because I have no correspondence nowadays to Petersburg and our booksellers here are very slow in ordering new scripts.

I have therefore been especially concerned about the atmosphere of the Moon, in how far it can have an influence upon the astronomical observations, e.g. on the duration of a solar eclipse or an occultation and the like; and in this view it is much the same thing whether one says that the atmosphere may be very subtle or may not be there at all since, as I at least have been made to believe, one does not detect in these phenomena the effect of the atmosphere. Whence, therefore, the extension of the Sun's diameter, which you have observed, may have arisen; whether it originates from a refraction of the lunar atmosphere or from the bending of the rays that also occurs in the case of other bodies which in our actual understanding have no atmosphere, or from something else, perhaps from the quality of the glasses of the [telescopic] tube through which the Sun's disc was projected, I am unable to distinguish. It remains to investigate whether the Sun's picture in a *camera obscura* appears widened in a like manner when, following Mr de l'Isle's instruction,[8] an artificial annular eclipse will be introduced. At least one is able in this way to obtain a clear impression of the annulus at a total eclipse and thus show that it is not produced by the lunar atmosphere. Meanwhile I am not willing to write anything regarding this matter neither generally nor even particularly against Mr Mylius;[9] but, since I am no friend of strife, as soon as someone in particular starts it, there for my part will the quarrel rest unless other circumstances make me feel strongly inclined [to do] the opposite.

I remain with complete respect,

Your obedient servant,

T. Mayer

* * *

Formula for the astronomical refraction[10]

Refraction in seconds =

$$70\tfrac{1}{3}B\cos\alpha e^{-3t/420}\left[\sqrt{(1+[16\tfrac{1}{2}e^{-t/420}\sin\alpha]^2} - 16\tfrac{1}{2}e^{-t/420}\sin\alpha)\right]$$

B denotes the height of the barometer in Paris inches

t = the degrees of heat measured upwards from the freezing point on the Réaumur thermometer [i.e. the temperature in degrees Réaumur].

e = the number whose hyperbolic logarithm = 1

Horizontal refraction = $70\tfrac{1}{3}Be^{-3t/420}$

α = apparent altitude.

This formula is in very exact agreement with observations. I have compared it with the experiments of Mr le Monnier at Paris[11] and of Mr Bouguer in the torrid zone,[12] and noticed no appreciable error, notwithstanding [the fact that] the coefficients $70\tfrac{1}{3}$, 420, $16\tfrac{1}{2}$ on which most depends, were determined for the most part through the crude theory.

6. *Euler to Mayer; Berlin, 18 March 1752*

Right Honourable Sir,
Most Highly Esteemed Professor

Your nice discoveries on refraction have made me take up this matter anew. Therefore let C be the centre of the Earth, O the position of the observer, CO = a, Z the zenith, SMO the refracted radius, OF its tangent at O, the

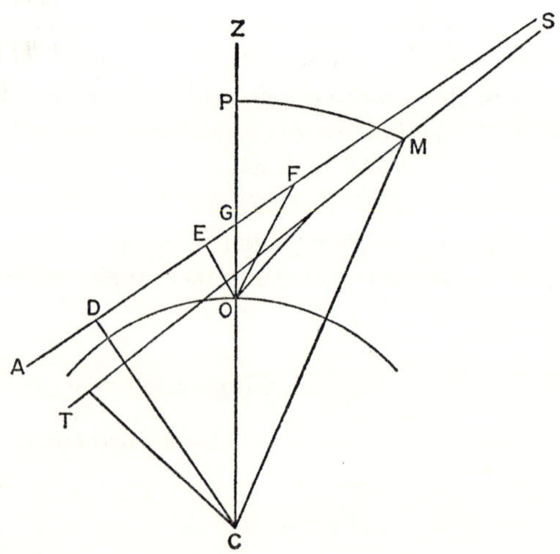

angle ZOF $= \zeta$, which represents the apparent distance of a star from the zenith.

Let γ be a fixed density of the air such that when a ray enters this air from a vacuum, the sine of [the angle of] incidence to the sine of [the angle of] refraction is equal to $1:\alpha$. When therefore the density of the air is equal to q, this ratio becomes $1:\alpha^{-q/\gamma}$, and consequently when a ray from the air whose density is q enters an air whose density is r, the sine of incidence to the sine of refraction becomes as $\alpha^{-q/\gamma}$ to $\alpha^{-r/\gamma}$.

Now let the density of the air at O $= k$, in M, however, or in $P = q$, having put the distance CM $=$ CP $= x$, so that q will be a particular function of x, such that $q = k$ when $x = a$ and $q = 0$ when $x = \infty$. The angle OCM is now put $= \varphi$, thus from the nature of refraction this equation

$$d\varphi = \frac{a\alpha^{(q-k)}\,dx \sin \zeta}{x\sqrt{xx - aa\alpha^{(2q-2k)}\sin^2 \zeta}}$$

follows. Or when a perpendicular CT is drawn from C to the tangent at M, CT $= a\alpha^{(q-k)} \sin \zeta$; and when the point M is infinitely distant, since then the tangent SD becomes the asymptote of the curve, the perpendicular CD $= a\alpha^{-k/\gamma} \sin \zeta$; and the angle OFG gives the refraction, when namely the infinite distance of a star is taken. It appears from that very equation, that the radius of curvature of the curve at M $= \dfrac{\gamma\alpha^{(k-q)}\,x\,dx}{a\,dq\,l\alpha \sin \zeta}$ and thus the radius of curvature at O $= \dfrac{\gamma\,dx}{dq\,l\alpha \sin \zeta}$. However, in order to establish the nature of the function q or x, one must take account of the hotness as much as the elasticity of the atmosphere. The latter is shown by the height of the barometer; the former, however, by the thermometer, but I assume that the thermometer is graduated in such a way that when the barometric height is one and the same [i.e. constant] the degree indicated by the thermometer is in reciprocal proportion to the density of the air.[1] Such a construction of the thermometer seems to be most in conformity to Nature, but it will not be difficult to reduce other arrangements to it.

I now consider a fixed temperature $= c$, which I give to the air with density $= \gamma$ for which the ratio of refraction approaches $1:\alpha$. This atmosphere will then have a particular elasticity, which ought to be shown through the barometric height $= b$. Thus b, c, γ and α are fixed quantities, with which I compare every other state of the air. Further, let the density of mercury $= n\gamma$, so that n will also be a fixed number.

This now being postulated, let the density of air at O $= k$, the barometric height $= h$ and the temperature $= g$; let the density of air in M or P, however, $= q$, the barometric height $= u$ and the temperature $= v$, thus $\dfrac{b}{\gamma c} = \dfrac{h}{gk} = \dfrac{u}{vq}$ or $\dfrac{k}{\gamma} = \dfrac{ch}{bg}$ and $\dfrac{q}{\gamma} = \dfrac{cu}{bv}$.

Since now the elasticity is proportional to the weight of the surrounding atmosphere, so must $q\,\mathrm{d}x = -n\gamma\,\mathrm{d}u$. Since now $\gamma u = \frac{b}{c}qv$, $q\,\mathrm{d}x = -\frac{nb}{c}(q\,\mathrm{d}v + v\,\mathrm{d}q)$, from which the relationship between q and x is determined, namely when the temperature v for every height in the atmosphere is known. Thus it also comes about that one may know how the temperature at every height, which can be very variable, is obtained. It seems, however, that the temperature always decreases with height. Now, when one assumes that at a height f above O the temperature is only half as large, one would not be much in error. When for the temperature in M or P at the postulated altitude OP $= x - a = z$, one assumes $v = \frac{fg}{f+z}$, and since $\mathrm{d}x = \mathrm{d}z$, one thence obtains $q = \frac{k(f+z)}{f} \cdot e^{-cz(2f+z)/2nbfg}$; consequently,

$$qv = gk\,e^{-cz(2f+z)/2nbfg}$$

from which one obtains the barometric height at P, $u = \frac{bgk}{\gamma c}e^{-cz(2f+z)/2nbfg}$. Hence, further: $\frac{\mathrm{d}x}{\mathrm{d}q} = \frac{\mathrm{d}z}{\mathrm{d}q} = \frac{-nbfg(f+z)}{c(f+z)^2q - nbfgq}$ and consequently at O $\frac{\mathrm{d}x}{\mathrm{d}q} = \frac{-nbfg}{(cf-nbg)k}$. Thus the radius of curvature of the curve at O becomes $\frac{-nb\gamma fg}{(cf-nbg)kl\alpha\sin\zeta} = \frac{-nbbfgg}{(cf-nbg)chl\alpha\sin\zeta}$. Now, we first of all know the radius of curvature of the curve at O, and in addition the inclination of the curve to the vertical at O, namely the angle ZOF $= \zeta$; thirdly, it is also known that the curve has an asymptote SD, whose distance from the centre of the Earth is CD $= a\alpha^{-k/\gamma}\sin\zeta = a\alpha^{\frac{-ch}{bg}}\sin\zeta$; and finally, the curvature of the radius SMO is certainly scarcely appreciable. Thus the refracted radius can be regarded as a portion of a hyperbola almost infinitely remote, like a convergent asymptote; and in this way, in my opinion, the true curvature of the radius will be most conveniently determined.

Hence let OE be a normal drawn to the asymptote, and AE the abscissa. One takes AE $= t$, OE $= y$, so for all possible curves this equation $\mathrm{d}t = -\mathrm{d}y\left(\frac{E}{y^m} + \frac{F}{y^n} + \frac{G}{y^p} + \text{etc.}\right)$ will occur, but since y is so very small, one can discard all higher powers of y itself and retain only the lowest, so that $\mathrm{d}t = \frac{-E\,\mathrm{d}y}{y^m}$. Since the curve has an asymptote, m cannot be smaller than 1. I now put the refraction or the angle OFE $= \theta$, so OGE $= \zeta + \theta$. Further, in order to abbreviate, let $\alpha = 1 - \mu$ and $\frac{nb}{a} = \nu$ and $f = \frac{1}{\delta}nb = \frac{\nu}{\delta}a$, so, since μ is so very small, $l\alpha = -\mu$, and $\alpha^{-k/\gamma} = 1 + \frac{\mu ch}{bg}$. From this the following determinations for this hyperbolic curve are found.

I. Since the angle OFE $= \theta$, so $\tan\theta = -\frac{dy}{dt} = \frac{y^m}{E}$

II. The perpendicular CD $= a\left(1 + \frac{\mu ch}{bg}\right)\sin\zeta = y + a\sin(\zeta + \theta)$

III. The radius of curvature at O $= \dfrac{vabg}{\mu ch(c/g - \delta)\sin\zeta} = \dfrac{(EE + y^{2m})^{3/2}\zeta}{mEy^{2m}}$

$$= \frac{y\sec^3\theta}{m\tan^2\theta} = \frac{y}{m\sin^2\theta\cos\theta}$$

Now since θ is very small, $\sin^2\theta\cos\theta = \theta\theta$ and $\sin(\zeta + \theta) = \sin\zeta - \frac{1}{2}\theta\theta\sin\zeta + \theta\cos\zeta$ and when one puts $\dfrac{vbg}{\mu ch\left(\frac{c}{g} - \delta\right)} = D$, one obtains

$$(2mD - \sin^2\zeta)\theta\theta + 2\theta\sin\zeta\cos\zeta = \frac{2\mu ch}{bg}\sin^2\zeta$$

consequently

$$\theta = \sin\zeta\,\frac{\sqrt{\cos^2\zeta + \frac{2\mu ch}{bg}(2mD - \sin^2\zeta)} - \cos\zeta}{2mD - \sin^2\zeta}$$

which formula is not very different from that which you found. It follows from this that when the term $\frac{2\mu ch}{bg}(2mD - \sin^2\zeta)$ is not so much smaller than $\cos^2\zeta$, which occurs so long as the angle ζ does not come too near to 90°, then the refraction will be $\theta = \frac{\mu ch}{bg}\tan\zeta$, in which the exponent m does not come into consideration. The horizontal refraction, however, when $\zeta = 90°$, becomes $= \sqrt{\dfrac{2\mu ch}{bg(2mD - 1)}}$ and is consequently dependent upon the exponent m. In order to determine this, I have sought for the radius of curvature from the evolute of the curve OM and after this still determined the hyperbolic line. Since I found $m < 1$, and since such [a result] is impossible, I therefore infer from it that one must assume $m = 1$* so that the curve OM will be a logarithmic [one], and then the horizontal refraction

comes out $= \dfrac{ch}{bg}\sqrt{\dfrac{2\mu\mu\left(\frac{c}{g} - \delta\right)}{2v - \frac{\mu ch}{bg}\left(\frac{c}{g} - \delta\right)}}$

* N.B. The exponent m ought to depend on ζ, or else be chosen in such a way that the horizontal refraction becomes larger.

Should the temperature be the same throughout the whole height of the atmosphere, then $\delta = 0$. Moreover, it is evident that the refraction depends upon the rise and fall of the barometer as much as of the thermometer. Everything therefore depends only upon the values of the letters μ, ν and δ, since it is then known than when b and c would signify an ordinary state of the air, μ is approximately equal to $\frac{1}{3300}$, and $\nu = 4\mu$; nb will amount to about a mile.

I am certainly of your opinion that the elasticity of the aether itself will also be reduced by the agitation of the particles, since such an agitation cannot arise without a certain union of the forces by which it is produced. Moreover, the theory still involves many inequalities due to the Moon, which cannot be properly determined on account of the greater difficulty of the calculations. Apart from those [inequalities] thus originating from the [non-spherical] shape of the Earth, and still just large enough to be taken into account, different [ones] which seem rather substantial also originate from the inclination of the lunar orbit to the ecliptic. As to these, however, they depend only upon the Moon's mean anomaly p, the Sun's mean anomaly s, and the mean distance of the Sun from the Moon ω; and as you have been kind enough to point out to me, these cannot easily be compared with my calculation, because I did not introduce into my calculation these elements associated with the mean motion. Mr Clairaut, however, whose article will be printed in Petersburg together with mine,[2] although it is still not accomplished, has done so. His formulae for the true longitude of the Moon are, according to your elements, as follows.

True longitude = mean longitudes	According to you [Mayer]
$-22797'' \sin p$	$-22736'' \sin p$
$+ 801 \sin 2p$	$+ 772 \sin 2p$
$- 37 \sin 3p$	$- 32 \sin 3p$
$+ 608 \sin s$	$+ 680 \sin s$
$- 21 \sin 2s$	$- 10 \sin 2s$
$- 113 \sin \omega$	$- 106 \sin \omega$
$+ 2394 \sin 2\omega$	$+ 2290 \sin 2\omega$
$+ 27 \sin 4\omega$	$+ 17 \sin 4\omega$
$+ 136 \sin (p - s)$	$+ 114 \sin (p - s)$
$- 102 \sin (p + s)$	$- 102 \sin (p + s)$
$- 12 \sin (2p - s)$	. . . .
$- 4606 \sin (2\omega - p)$	$- 4540 \sin (2\omega - p)$
$+ 43 \sin (4\omega - 2p)$	$+ 33 \sin (4\omega - 2p)$
. . . .	$- 36 \sin (4\omega - 3p)$
$- 68 \sin (4\omega - p)$	$- 43 \sin (4\omega - p)$
$- 199 \sin (2\omega + p)$	$- 104 \sin (2\omega + p)$
$+ 40 \sin (\omega - p)$	$- 37 \sin (\omega - p)$

True longitude = mean longitudes	According to you [Mayer]
$+ 134 \sin(2\omega - 2p)$	$+ 200 \sin(2\omega - 2p)$
$+ 7 \sin(\omega + p)$	. . . .
$+ 9 \sin(2\omega + 2p)$	. . . .
$- 18 \sin(2\omega - 3p)$	$- 60 \sin(2\omega - 3p)$
$+ 21 \sin(2\omega + s)$	$+ 23 \sin(2\omega + s)$
$- 162 \sin(2\omega - s)$	$- 10 \sin(2\omega - s)$
$- 44 \sin(2\omega - p + s)$	$- 55 \sin(2\omega - p + s)$
$+ 202 \sin(2\omega - p - s)$	$+ 271 \sin(2\omega - p - s)$
$+ 22 \sin(2\omega - 2p + s)$	$+ 88 \sin(2\omega - 2p + s)$
$+ 29 \sin(2\omega + p - s)$	. . . .
$- 12 \sin(2\omega - 2p - s)$	. . . .
$- 81 \sin 2u$	
$+ 90 \sin(2\omega - p + 2u)$	where u denotes the mean dis-
$+ 72 \sin(2\omega - 2p + 2u)$	tance of the sun from the node ☊.

From this comparison you will perceive how far your determinations are in agreement with those of Mr Clairaut; meanwhile, this gentleman has written to me that he has since that time made marked changes in his [formulae]. In order to determine the motion of the apogee correctly it is not necessary to consider the eccentricity as variable, but actually only to introduce the term $2\omega - p$ into the calculation, as both Mr Clairaut and I have done, despite our having approached the problem from quite different directions.

With regard to the Moon's atmosphere you should on no account withhold your ideas, as far as I am concerned; I have only written about the phenomenon at annular eclipses[3] and my views had to be adduced from the cause itself, in which, since only a single phenomenon is involved, I might easily have erred. To this end, I have proposed other observations concerning the occultations of fixed stars, in order to prove: does the Moon actually have a subtle atmosphere or not? and the same observations ought to decide my opinion on the whole controversy. Since an excellent micrometer is now available for this purpose, no one is more fit to discuss this matter than yourself.

I have the honour to remain with the most complete respect

Your most humble servant,

L. Euler

May I most humbly beg you to transmit the enclosure to Cassel, since I believe that this can be done without expense.

7. *Mayer to Euler; Göttingen, 24 April 1752*

Dear Sir,
Honourable Professor.

As far as I can judge you have solved the problem of the astronomical refractions so conveniently and generally that my troubles seem to me as nothing in comparison. I have compared your newly-found formula with observations, and found that it gives the refractions from approximately the 23rd up to the 90th degree of the altitude as accurately as one may wish. However, if the altitude is less than 3 degrees, the corresponding refraction is invariably somewhat too small, this difference being most pronounced in the special case of horizontal refraction. My formula, which I have previously applied, also has this same error. The error is, however, a little less in your formula than in mine, and depends principally upon the exponent m; m ought to be equal to 2 for the horizontal refraction, if the formula is to agree with the observations; however, when the altitude is about 5 to 10 degrees, m should not be greater than unity. But as you yourself have pointed out, m has no particularly marked influence at greater altitudes; and it can therefore be either $=2$ or $=1$. Consequently, it appears that the exponent m must depend either upon the altitude or on the apparent distance from the zenith, and thus not be constant, or the equation for the refraction of the solar rays must not be $\mathrm{d}t = -\mathrm{d}y\left(\frac{\varepsilon}{y^m}\right)$ but rather $\mathrm{d}t = -\mathrm{d}y\left(\frac{\varepsilon}{y^m} + \frac{F}{y^n}\right)$ or alternatively $\mathrm{d}t = -\mathrm{d}y\left(\frac{\varepsilon}{y} + \frac{F}{y^2}\right)$ in which ε is very much smaller than F at low altitudes, but $\varepsilon > F$ at higher ones. Since, however, yet another undetermined quantity is hereby added to the foregoing, a fourth condition will still be required in addition to the three that you have proposed for the determination of the curvature of the light beam. Perhaps one derives these if one assumes that the direction of the sunlight has two asymptotes which also the nature of this light path certainly seems to permit. At least it is certain that two such asymptotes actually exist for the ray which touches the horizon.

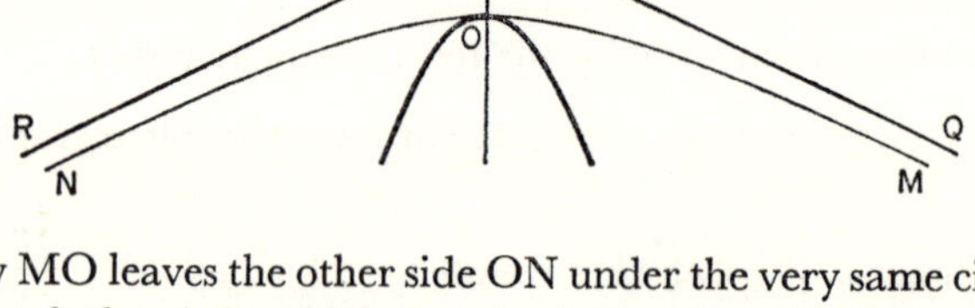

Such a ray MO leaves the other side ON under the very same circumstances again through the air in which it arrived; it suffers the same refraction and therefore has the double asymptotes GQ and GR. This has made me assume for the ray NOM a common hyperbola not only in the horizon but also at greater heights. On account of the great difficulties, however, I have not yet completed the calculations.

Some time ago, and by reason of the treatise "De usu machinarum maxime lucroso" which you have had printed in part X of the *Comment. Petrop.*,[1] I wanted to investigate the motion of a so-called overshot water-wheel: like the kind generally used in conjunction with mining machines. I have also brought this investigation completely to its close in the case when the load on the axle of the wheel will be raised as it were, by means of a cord, where its distance from the hypomochlius is consequently always unchanged. Only if I suppose that the load be applied at a crank, a case which actually occurs with the hypothetical artificial fountains, I certainly see that the motion of the wheel must be very irregular, but the solution or the determination of the speed itself appears so difficult to me that I have again left it completely alone. Meanwhile the first case has nevertheless given me the opportunity of [making] various notes which do not allow me to regret my wasted efforts.[2] The main thing as much with this type of machine as with most others is this, that the force which ought to put the machine into motion is variable and not continuous, as one is usually accustomed to assuming in ordinary mechanics. The motive force (e.g. the water, in the case of the overshot water-wheel) is in fact mostly weaker if the machine moves more quickly; and since here again the speed depends upon the force, the investigation of such a motion in this way will be extremely difficult. It is this very fact that leads me to believe that the motion of most machines and especially those which are moved by means of water, will ultimately become uniform. The theory of machines can nowhere else await an improvement with greater hope than from you.

I have the honour to remain with the most complete respect,

Your obedient servant,

T. Mayer

8. *Mayer to Euler; Göttingen, September/December 1752*

I learn with very great pleasure through your consideration and further explanations, that you have set great store upon the subject of machines.[1] Since then, on account of other obligations, I could not continue my previously-mentioned effort in this matter; however, the difficulties which then forced me to abandon the investigation of the motion of the overshot water-wheel if the load is applied at a crank, are all still known to me. This type of motion is very different from that which an undershot wheel has in just such conditions and, as to the latter, I think that the determination of the same is not nearly as difficult as in the former case. I suppose, in the script which you thought will soon appear in the Memoirs,[2] that the particular circumstances which apply to overshot wheels will not be concerned, so I am therefore taking the liberty of mentioning them at present. The major difficulty in this investigation is not actually caused by the crank but rather by the water falling upon the paddle-wheel, from which the load on the crank ought to be moved. If such a wheel has an irregular motion, it will soon be laden with more or with less water, because the inflow from the channel is not equal to the outflow from the paddle. The former is actually reciprocally proportional to the speed with which the paddles of the wheel rotate under the channel, whereas the latter would be directly proportional to the volume only if the paddles were each filled by the same amount of water; however, this is not the case.

9. *Mayer to Euler; Göttingen, September/December 1752*

Dear Sir,
Honourable Professor,

I have now investigated anew the formula for astronomical refraction which you had the goodness to communicate to me, and have varied in all possible ways the value of those letters on which the temperatures in different layers depend, in order to see whether it may be possible to obtain a more accurate agreement of the formula with the observations. Yet it has given me no satisfaction. According to the observations of M. Cassini which are in the Memoires for 1714,[1] the horizontal refraction at average temperatures is 38 to 40 minutes [of arc], but at colder temperatures fully 50 to 52 minutes. However, already on the 1st and 2nd degrees of altitude it decreases so rapidly that it seems to me impossible to introduce this decrease into the formula, without causing the refractions at heights of 5° to 10° and still further to be much too small. I am completely disregarding the horizontal refraction, however, so it will still suffice to determine by means of the formula all the refractions from the 2nd to the 90th degree of altitude as accurately as is desirable; which can also generally be done through my [own] formula. Yet there would certainly appear to be an unknown cause which makes the horizontal refraction so large; or perhaps it lies in the fact that in the theory the formula was not applicable on the double asymptote of the radius, as I have already mentioned previously. One can be content in the meantime with the present [formula], firstly since only very few stars apart from the Sun and Moon can actually be observed at the horizon; and in addition, since at such low levels changes are occurring almost instantaneously in the refraction, as I have often perceived on the image of the rising or setting Sun. Besides, it seems to me that the refraction table in the Berlin astronomical calendar and which has Newton as its originator, as the Transactions show, is in more accurate agreement with the law which the observations provide; I merely do not know upon what theory it was based.

I am very covetous of the writing on the theory of the Moon both of yourself[1] and of Mr Clairaut,[3] which I shall now soon obtain along with the new tome of the memoirs of the [Berlin] academy of sciences,[4] through the bookseller from the exhibition. It is one of our observatory's most noble intentions to make as uninterrupted a series of lunar observations as possible, and with it to come to the aid of the theory. With a favourable summer both the building and the instruments will be so far developed that we can then begin to observe.[5]

At the last meeting of our local scientific society I have read the report of an instrument which is only 2 inches in radius but is so exact that every terrestrial angle of more than 10″ to 15″ can be accurately measured with it, and which therefore accomplishes as much as a 3 ft radius quadrant.[6] This advantage is not obtained through some micrometer as is usually the case, but rests principally upon one's obtaining a certain multiple (e.g. ten or twenty times) of the angle as a result of repeated operations; by means of which the error in the simple angle becomes so much smaller. It would be desirable if this practice were also to be used in astronomical observations which, however, owing to the scintillation of the heavenly objects, certainly does not seem possible. I will nevertheless test it by meridian altitudes, which do not change very quickly, as soon as the artisan will have completed the necessary instruments for that purpose.[7]

The practical parts of mathematics, which I am obliged to teach at this academy, often give me the opportunity of thinking about the connection of the theory with practice. Almost everywhere I profit from your contributions to these sciences. Mechanics, astronomy, music, artillery, optics, navigation owe infinitely much to you. Architecture still awaits this good fortune. Everything which the architects speak about concerning the stability of a building is still so indeterminate and doubtful, that one should judge it as nothing against that which a correct application of mechanics to architecture could in future provide. I have formulated for myself many relevant propositions which I keep trying to solve time and again; I have already solved a few which pave the way for the others. Below are the following:

1. Given the weight which a horizontally-lying beam can bear if it is completely supported at both ends and the weight is brought to the middle; the question is, how much can it bear if the weight were suspended at any other given place apart from the middle? Similarly, if the beam lies obliquely.
2. If one knows how much weight such a beam can support at its centre without breaking; to find how heavily it could be laden if the load is distributed equally or unequally through its entire length.
3. To find the optimum shape of a corbel AB which is rigidly attached to the wall CD, and which has to carry a load E at A, given that the whole length AB of the uniformly strong corbel may always be kept from breaking; or conversely, if the shape of the corbel is known, to find at which place it must break, if the weight E together with the stone's own weight is too great.

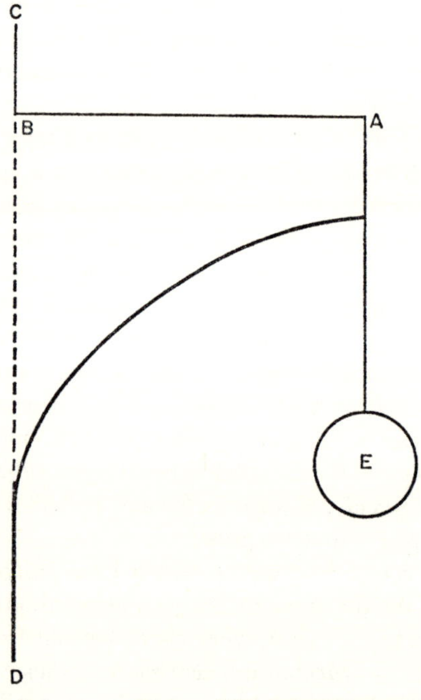

In the solution of these problems, I have made use of the rule that the strength of a uniformly thick beam or stone varies directly as the square of the height, and as the breadth; but inversely as the length. Since the experiments of Musschenbroeck[8] and Buffon[9] do not entirely agree with this rule it would be desirable if the strength of the wood, etc., would be determined from a better theory than the Galilean one is.[10]

10. *Mayer to Euler; Göttingen, 7 January 1753*

Dear Sir,
Honourable Professor.

I do not know how you have taken my slowness in answering, since the delay has lasted for almost half a year.[1] I had nothing in particular to write about that was worthy of your attention; on the other hand I waited for the new volume of the memoirs of the Royal Academy there [i.e. in Berlin], containing your and M. Clairaut's theory of the ☽, through the local bookseller. I have not yet received the latter writings. In the former I noted with great pleasure the investigation of Prof. Segner's new machine.[2] This gave me the opportunity of making a comparison between this machine and the common overshot water-wheels, and I have found that the effect with both amounts to almost the same thing. You have shown that Segner's machine loses so much less of its effect, the faster it rotates. With the overshot water-wheel, however, one has nothing to say about this; on the contrary another circumstance exists which causes such a wheel never to provide its full effect. This depends upon the shape of the paddles as these can never be made so that they do not lose a portion of their water before they reach their lowest position. Nevertheless, one can arrange that no more than about the 40th or 50th part of the whole force of the water is lost merely by giving the paddles a somewhat different shape than the common one; whereas, with the usual construction, usually a 4th part of the force is lost.

I shall undoubtedly obtain the writings of the Moon's motion through the bookseller, from the New Year's exhibition. Up to now I have examined these matters once more, but could again not derive the total [observed] motion of the apogee. Against this, I now have the equations of longitude as accurately as I have wished. There are no more than 13 of them, which I have arranged in tables, and through which I shall be able to calculate the longitude of the Moon so correctly, that the error never amounts to 2 minutes. From a total of more than 200 observations which I have compared with these tables, there are only about 10 or 12 which deviate by about 1′ 30″, yet not one varies by a greater quantity. The reason why I have so few equations lies in the application itself, which you can understand from the following.

Note. Here s is the mean anomaly of the ☉, p the mean anomaly of the ☾, ω the [angular] distance of the mean ☽ from the true ☉, δ the mean distance of the ☽ from ☊.

Equations of the longitude of the ☽	Equations of the mean anomaly of the ☽	The equation of ☊
(I) $+11'\,20''$ sin s -10 sin $2s$	(A) $+20'\,36''$ sin s -20 sin $2s$	(B) $+10'\,18''$ sin s -10 sin $2s$
(II) $+\ 3'\,45''$ sin $(2\omega-2p)$ $+28$ sin $(\omega-p)$	the same	
(III) $-\ 0'\,54''$ sin $(2\omega+s)$		
(IV) $-\ 1'\ \ 2''$ sin $(2\omega-s)$		
(V) $+\ 1'\,48''$ sin $(2\omega-p+s)$		
(VI) $+\ 1'\,12''$ sin $(2\omega-p-s)$		
(VII) $+\ 1'\,30''$ sin $(2\omega+p)$		
(VIII) $+\ 0'\,58''$ sin $(2\delta-p)$		
(IX) $+\ 0'\,40''$ sin $(p-s)$		
(X) $+\ 0'\,47''$ sin (distance of ☊ from $\odot$ squared)		

If these X equations of the ☽'s mean longitude, together with the equation (A) of the mean anomaly, are applied, one again tries to find the distance of the ☽ from the $\odot$, which may be $= \bar{\varpi}$; moreover, should the ☽'s anomaly corrected through equations (X) and (A) be $= \bar{p}$, then one still has the following equations

$$\text{(XI)}\begin{cases} -6^\circ\ 18'\ 20''\ \sin \bar{p} \\ +\qquad 13'\ \ 0''\ \sin 2\bar{p} \\ -\qquad\qquad 36''\ \sin 3\bar{p} \end{cases} \qquad \text{(XII)}\begin{cases} -1^\circ\ 20'\ 42''\ \sin\,(2\bar{\varpi}-\bar{p}) \\ +\qquad\qquad 35''\ \sin\,(4\bar{\varpi}-2\bar{p}) \end{cases}$$

If these two equations are also applied, one sets the resulting distance of the ☽ from the $\odot$ $= \bar{\bar{\varpi}}$, then the variation is

$$\text{(XIII)}\begin{cases} +40'\ 21''\ \sin 2\bar{\bar{\varpi}} \\ -\ \ 1'\ 56''\ \sin \bar{\bar{\varpi}} \\ +\ \ 0'\ \ 2''\ \sin 3\bar{\bar{\varpi}} \\ +\ \ 0'\ 17''\ \sin 4\bar{\bar{\varpi}} \end{cases}$$

Through these 13 equations the true place of the moon in its orbit is determined. The reduction is always $-6'\,57''$ sin $2\bar{\delta}$, where $\bar{\delta}$ denotes the distance of the place of the true ☽ from ☊ corrected by equation B. The latitude of the true ☽ can be found very simply in the following way. Let $\bar{\delta}$ be, as before, the distance of the true ☽ in its orbit from ☊ corrected by equation B, and $\bar{\bar{\varpi}}$ the true distance of the ☽ from the $\odot$; therefore the northern latitude of the true ☽ $= +\sin\,(5^\circ\,9'\,0'')\,\sin\bar{\delta} + 9'\,30''\,\sin\,(2\bar{\bar{\varpi}}-\bar{\delta})$. In such a manner I do not require to find the true longitude of the node nor even the inclination to the ecliptic, through which the calculation of the latitude would otherwise be very exasperating. The calculation according to these principles is certainly much more convenient than if one were to adopt another classification in the application of the equations, as I previously used to think; since I have used the mean longitude of the Sun and Moon as well as the approximate mean anomaly. Should these calculations be in agreement with the observations, the following results, which I have established from a comparison of the observations of this and the previous century, must also be taken into consideration.

	s	°	′	″
Epoch of the longitude of the mean ☽ on the meridian which precedes 1 Jan. 1744	3ˢ	28°	48′	2″
Mean time with respect to the Paris meridian				
Epoch of the lunar apogee at the same time	9	16	51	8
Epoch of ☊	1	6	20	40
Motion of the ☽'s longitude in 60 Julian years	1	10	43	20
Motion of the ☽'s apogee	9	11	26	0
Motion of ☊ from Flamsteed's tables[3]	2	20	30	45

Since all other tables fix the motion of the Moon's longitude as smaller than I have found, this is a very strong argument for the postulate that the Moon's motion is always getting faster. I have [taken] the motion of the apogee [to be] markedly smaller than [in] other tables; [a supposition] which also appears likely to come into agreement with the theory.

I must still remark that most of the above-mentioned equations are not as I found them from the theory. The comparison of the observations with my figures has necessarily caused me to make several small changes in it [i.e. the theory] which are, however, not so appreciable that one ought to believe that Newton's theory of gravity were false. I have completely and utterly ignored many small equations, which are similar to those which depend on the shape of the Earth, since they scarcely amount to 12″ or 15″.

I have the honour to remain with the greatest possible respect

Your obedient servant,

T. Mayer

11. *Euler to Mayer; Berlin, 24 April 1753*

Before everything, I most humbly beg your forgiveness that I have been so late in answering your friendly letter filled with the most important discoveries. Various duties with which I am overwhelmed so often bring me out of communication with such profound researches that I do not have confidence to reply to them. You have certainly greatly improved the Moon's motion beyond what was previously known, and as your tables of the Moon's position show an accuracy of 2′, a more exact determination is scarcely to be hoped for, since it is for the most part uncertain whether this error does not largely depend upon the observations themselves. Concerning the Moon's apogee, whose motion you have still not estimated with sufficient accuracy, the principal question is whether this is found to be significantly greater than 0° per year; then, since the difference arises from some lunar inequalities, it is simple to consider that the discrepancy originates from the remaining [ones]. Meanwhile you will have obtained M. Clairaut's work, and I have also already completed my treatise on the same subject some time ago: however, because I included so many other copies which the [Berlin] Royal Academy from time to time order to be sent out, I doubt whether such a dispatch from Leipzig will still occur before the exhibition. I candidly admit to you in advance, however, that my theoretically deduced tables often differ by over 5′ from the truth. I have strongly recommended your profound insights into the same important research both to our president[1] and to others in accordance with its worth, and have testified to the truth that no one in astronomy has improved it, or will improve it, more than yourself. Whereupon, I have been ordered to sound out whether you would not want to resolve to take up a proper appointment here in Berlin. From what I have heard, your conditions in Göttingen are not so advantageous that I should not trust myself to obtain better for you here[2] and since I also very much doubt that your splendid talent will be financially recognised in Göttingen, I am assured that such a positive decision would be not only to your advantage, but also be quite a special pleasure [for you]. It would especially be the most sensitive pleasure for me to experience such an agreeable association with a man of such profound insight. Above all, however, please keep this matter as secret as possible, and let me know frankly whether you could resolve upon such a change or not? Moreover, [let me know] under what conditions you were engaged at Göttingen, so that I may be able to work out something better here. You can safely divulge your views on this to me, and be assured that I shall not make use of them other than to your advantage.

In anticipation of a favourable reply to this, I have the honour to remain with the greatest possible respect.

Yours, etc

L. Euler

12. *Mayer to Euler; Göttingen, 7 May 1753*

Dear Sir,
Honourable Professor

Your last letter, which I received on the 5th of this month, contains so much of importance for me, that I find it necessary to reply to it without much delay. To start with, I know that I am under an obligation, sir, to return you a very special thanks for such a good—and from my side, undeserved—recommendation [as has been] conferred upon me in Berlin, more especially since this, and not my feeble efforts regarding science, is the actual reason for such an advantageous proposal for me. Although I plainly have little cause to be discontented with my present circumstances in Göttingen, I certainly want to make up my mind to go to Berlin, where I would find pleasure in contact with so many great scholars, and especially with yourself, which is the one thing that I still rather miss here and which I value far more highly than all other kinds of improvements of my temporary circumstances. Since, however, one is certainly unwilling to suffer through doing this, and [since] you want to know explicitly my conditions here, I can openly reveal them. I am here as Ordinary Professor of Economy under an annual basic salary of 450 Rthl. If I reckon in addition what our scientific society, likewise the college, together with other incidentals bring in, I increase it to about 600 to 800 Rthl. per annum, which is not just too much for the cost of living here (I do not know how it is in Berlin). Meanwhile it could happen that in the case of [my being called to] another appointment I might receive an increase of salary, though it is not at all likely to tend to occur otherwise; and on balance I certainly want to move to Berlin. Together with Mr Segner, I have here to look after the observatory which will now soon be completed, and I am not eager to let this advantage out of my hands. Yet because I know that the observatory there [i.e. in Berlin] is likewise in a useful state, I hope not to make the matter any more difficult on this account. In general one is completely free here from all civic duties and worries. You see here the frank, true state of the situation. I hope that you are not tempted to consider it to be too advantageous for me. Owing to certain circumstances,[1] it would be very convenient for me to receive your more detailed explanation of this soon.

My tables on the Moon's motion are now printed in Volume II of our Transactions,[2] although I have cause to complain about the bad quality of the printing regarding the arrangement of the tables. By means of several minor alterations, which I have first made after the time when I had the honour to give you news of the matter, I hope to have brought it to the state that the error is still scarcely 1′. Since then, a total of 139 observations of the Moon, dating from September 1743 up to April 1745, have also been put into my hands by a man called Schumacher,[3] who travelled through here a few months ago and who may be not unknown to you; from these I have ascertained with no little pleasure that they agree on the whole with the tables to within 1′, with one exception; namely that of the 3 December 1744, which disagreed by about 1′ 31″ and which is perhaps rather unsatisfactory, as it is certainly difficult to determine the position of the Moon by observations more accurately than to 1′. I have compared with the tables most solar and lunar eclipses which were observed after the beginning of the last century—particularly since the invention of telescopes and pendulum-clocks[4]—and never found a greater difference than 1′ 10″, which appears to be mainly due to a yet-unknown variation of the eccentricity of the Sun. Those longitudes of the ☽ which Dunthorne quoted in the [Philosophical] Transactions[5] out of the observations of Flamsteed[6] in support of the correctness of his tables, and which often differ from them by 3′ to 4′, agree on the contrary with mine up to 1′ or 1′ 30″; and from such observations as Flamsteed's are, which have often involved larger errors with the fixed stars, no better agreement is to be expected. I am leaving out of consideration other observations, e.g. those which appear in the Paris Memoirs, as well as my own, since the number of those in comparison to the others which I have compared with my tables, is very small. So much is certain, that the tables give the Moon's position as accurately as could hitherto have been obtained through observations, and that therefore no greater reliability in these can be supplied or hoped for until more diligence is first of all applied to the method of observing. I have quite successfully observed yesterday's transit of Mercury across the Sun from 5^h 4′ to its emersion which, incidentally, followed at 10^h $52\frac{1}{2}$′, despite the fact that it could not be accurately observed due to clouds, and I find that at the next conjunction with the centre of the Sun, Mercury will be around 6^h 56′ 15″ true time, in the morning, at a southerly distance of 2′ 26″ from the centre [of the Sun's disc]. The observed conjunction according to the longitude was about 7^h 3′ 20″, the observed latitude at that time being 3′ 30″ south. By subtracting the parallax, the position of true conjunction in the ecliptic is about 7^h 4′ 12″, the true longitude of Mercury at the time of conjunction 2′ 27″. I find from my solar tables that the longitude of the ☉ at this time is 15^h 47′ 54″, or rather, calculated from the mean equinox, 15^h 47′ 42″ ♈ and according to Halley's tables[7] the inclination to the ecliptic of Mercury's orbit is 6^h 59′ 20″, thus I find further that at about the above-mentioned time of the true conjunction 7^h 4′ 12″ the heliocentric longitude of Mercury [is] 15^h 47′ 42″ ♈, the heliocentric latitude 3′ $0\frac{1}{2}$″ south, the distance of Mercury from ☊ in consequence 24′ 44″, hence the longitude ☊ Mercury [is] 15^h 22′ 58″; the apparent diameter of Mercury was 11″ or at the most, 12″.

From a comparison of older observations with these, of whose reliability to a few seconds I am assured, I have found the annual motion of ☊ Mercury to be 42″ to 45″ and thus smaller than the precession of the equinoxes, in contrast with the annual motion of Mercury's aphelion 55″ to 56″; and hence larger than the precession of the equinoxes. This latter motion, which Halley estimated at $52\frac{1}{2}$′, but Cassini at 1′ 20″, is largely to blame for Cassini's tables on this occasion varying so much from the truth. This transit is therefore important for the theory of Mercury, because it is only the second which was observed at the descending node, whereas in contrast the other 8 all fall together at the ascending node.

I have been very unfortunate with the prize essay of M. Clairaut, since up to this hour I have not been able to obtain it. My bookseller has lingered so long that it was no longer to be had in Leipzig. I myself wrote there, but it could not be obtained for me. Nevertheless I hope to get it quite definitely at this exhibition.

I have deduced from my computations the daily motion of the Moon's mean anomaly as 13° 3′ 25″, which yields more than 40° for the annual motion of the apogee. However, since right at the beginning of the calculations I substituted figures instead of symbols, I am unable to say on what this motion actually depends, and I have too little time left to go over the whole of the calculation once again. Meanwhile I have no doubt that in your treatise, which I look forward to with great eagerness, you will have sorted out this matter so well that my efforts on it will surely have been superfluous.

I suppose that M. Clairaut will now also deduce the motion of Saturn and Jupiter. I regard this undertaking as much more difficult than the determination of the Moon's motion. Although the inequalities are very much smaller than in the case of the Moon, so is their number seemingly not much less. After having seen how you have tackled this problem regarding Saturn in your prize essay,[8] I have in much the same manner made a test with Jupiter; through which the difficulties with which these researches are bound up immediately became apparent. The equations which I obtained for the motion of Jupiter are not sufficient to predict the position itself more accurately than to 4′.

I have, sir, the honour to recommend myself and my circumstances to your benevolence and to remain with the greatest respect,

Your obedient servant,

T. Mayer

13. *Euler to Mayer; Berlin, 15 May 1753*

Right Honourable Sir,
Most Highly Esteemed Professor

I acknowledge with the most grateful thanks your candid explanation of my proposal, and wish no more than to receive soon such instructions about what can be done regarding your proposals, with which you may be completely satisfied. The power that our President left with me before his departure does not extend further than to offer you 550 Rthl. yearly, this being as much as the academic treasury is in a position to bear, and upon which one can certainly count. Nevertheless, as I have now heard, the cost of living is not so dear here as in Göttingen, and just as much can also be earned on the side through lectures. On top of this you can be given the definite assurance of receiving an increase during the first vacation, since His Royal Majesty has the intention of decreasing the number of paid members, and against this to make their salaries more considerable; thus I have written with today's post to our President in order that he can straightaway beg an increment for you from His Royal Majesty, which matter will present no difficulty. Meanwhile, you can bank upon the stated 550 Rthl., and if your circumstances permit you to be content with it, you may begin your journey rightaway. I do not, however, doubt [that] the consideration of your particular earnings and important discoveries, from which you always give me new examples, will within a short time still effect an increase of several hundred Rthl. Meanwhile I can assure you that you would live so pleasantly here that no one would concern himself with knowing your earnings. In particular, I could wish for nothing more than to have you here, as I can also render no greater service to the Academy than to assist you [in making] such a priceless achievement. The local observatory, which would be put entirely at your disposal, is in quite a good state, and on your recommendation everything which is still lacking could easily be put right.

I have the greatest reason, sir, to congratulate you sincerely on your so fortunately executed lunar tables, as I consider these to be the most important discovery which astronomy can hope for from analysis, and from which I myself have experienced more than enough difficulties. The 139 observations received from the so-called Schumacher came from me, as [they are] those which I myself elicited from Dr Bradley.[1]

We have also observed the transit of Mercury here, although we were actually much too often disturbed from our work by the frequent clouds. You have, however, acquired something very important from it, as you have so fortunately corrected the elements of Mercury's motion out of it. It will now be simple to prepare accurate tables of Mercury, from which only Halley's supposed eccentricity has no need of any correction, which you would easily find out, however.

Time requires me to break off, therefore I have the honour to remain with the greatest respect, sir,

Your most obedient servant,

L. Euler

14. *Mayer to Euler; Göttingen, 23 May 1753*

Dear Sir,
Honourable Professor

I acknowledge with the greatest indebtedness your very kind communication and provisions concerning my awaited fortune in Berlin. I find nothing of any importance to be criticised in the proposals made; also, in order not to forfeit the pleasure and the honour of your personal acquaintance, nor to try to make myself unworthy of this fortune, I shall not press the matter with too great a force. For all that, it is only fair to let you know that in case I should earnestly desire to get away from Göttingen, [I would hope that] a proportionate amount of the burden of the travelling expenses would be accorded to me. Besides this, I still cannot positively declare myself [for or against moving to Berlin] until I learn what President de Maupertuis replied to your very kind recommendation. To confess the truth, I would certainly not make my circumstances worse by [accepting] the offered 550 Rthl., yet I would not raise [them] much higher. Our scientific society has offered me 150 Rthl. for preparing an astronomical calendar, which I can likewise count as my normal income; thus, if it were not [for the fact] that the enormous effort [associated] with work of this nature is certainly not compensated by 150 Rthl., I am just as well remaining here. You could decide the matter through a small increase in salary, and this would please me more than the hope of a considerable betterment in the future. It is, moreover, unknown to me whether one has there a free home or equally valuable benefits. Similarly, what a member of the academy's customary and responsible duties are; also whether one is free there from the usual civic responsibilities, how much it is to rent a house, in case one must see to it oneself, etc. I wish no more than that the matter will come to a satisfactory conclusion, so that I should have you to thank not only for that which I thoroughly understand in the sciences, but also for my temporary good fortune and indeed for everything else.

You have such a great share in the theory of my lunar tables that I would be completely unrecognised if I had wanted to appropriate these for myself alone. I am sincerely of the opinion that I would perhaps not have progressed far in this by myself, if you had not, through your admirable writings, especially through the masterly treatise on the motions of Saturn and Jupiter,[1] paved the way for me. All that I have found out about this from others and what I myself have also previously attempted in it, is so incomplete and undertaken in such a manner that the true theory of the Moon would undoubtedly have still remained long hidden, had you not first reduced the major difficulties with incomparably better success.

Although I am certainly convinced that my tables seldom depart by over

1′ from the truth, I nevertheless still do not regard it as impossible to approximate still more closely to it if only there were first of all more precise observations, and [if] afterwards someone had so much patience and time to carry out the approximations in the theoretical calculations still further. The latter would indeed be very difficult in [the case of] the Moon's longitude. It appears to me to be quite simple, however, to formulate completely the still-lacking equations for the latitude. In my tables I have discarded all the inequalities in the latitude which depend upon the eccentricity of the ☽ and of the Sun, of which there is nevertheless quite a number, and some of which could mount to 30″ and perhaps beyond. I have carried out the calculation for these small inequalities in latitude three times, but on every occasion found them to be notably different, so that I can only believe that it may be wrong time and again through the numerous terms in an equation, or else where an error of calculation was incurred. Which is entirely possible in a task that was so often interrupted by others. Meanwhile, it nevertheless seems to be worth the trouble to determine the latitude of the Moon still more precisely, because a great advantage in the discovery of the geographical longitudes from the occultations of the stars and from solar eclipses would thereby be obtained.

In order to improve the tables of Mercury, I have already assembled a variety of material, and made a test to determine the eccentricity. I find, however, that provided that the observations which I have used are to be trusted, I could not have Halley's eccentricity instead; and that, moreover, the mean distance from Mercury to the Sun does not completely agree with Kepler's [third] law.[2] Thus all this requires a still more accurate investigation. Only it is not possible to attain certainty in this way from the transits of Mercury in front of the Sun. The observations which ought to be suitable for this purpose must be selected, and made under suitable circumstances; and it is difficult to discover such among the few which one acquires from astronomers. However, I am detaining you much too long with these trivialities. I therefore close and have the honour to remain with the greatest respect, sir,

Your obedient servant,

T. Mayer

* * *

I have given myself all possible trouble to obtain the new edition of the English Tables of Logarithms in quarto-size, 1748,[3] but in vain. I have been informed that you possess a spare copy of it. If you could get hold of one, I would most humbly beg that it be transmitted at your earliest convenience. I shall be most willing to pay the postage as well as the price.

15. *Euler to Mayer; Berlin, 13 June 1753*

Right Honourable Sir,
Most Highly Esteemed Professor

Your decision to accept the position of an astronomer at the local academy has given me quite an exceptional pleasure, and I wish very sincerely that the matter will soon be resolved to your advantage and pleasure. President von Maupertuis has also testified to his pleasure over it, and personally considers it to be highly probable that such a pension will immediately be made out to you, so that you will lose nothing by such a change. The pension that the academic treasury can give out is, as [previously] stated, 550 Rthl. yearly. We are, however, in no doubt that His Royal Majesty will assist the academy with an increment; regarding this, however, the return of the president must be awaited, then the travelling expenses will easily be made out straightaway. I only regret that this matter must still be somewhat postponed, yet I hope that it will turn out to your advantage.

I have never had a copy of the large logarithmic tables of Gardiner in 4to,[1] and that which I possess is a present from our president who had subscribed to it. The Honourable Fieldmarshall Schmettau[2] had several copies made from it, but all were distributed in the auction.

That you have found different values by the repeated calculation of several lunar inequalities is a circumstance that also tortured me enough. The reason for it is, that the denominator of the fraction which determines the coefficients is very small, and since it can be deduced only by approximation, the smallest error can produce an enormous difference.

Meanwhile, I have the honour to remain with the greatest possible respect,

Your most humble servant,

L. Euler

16. *Mayer to Euler; Göttingen, 15 July 1753*

Dear Sir,
Honourable Professor

I have been hindered until now by different affairs from answering your last very valuable letter. In consequence of it, I shall await meanwhile the president's return, in the hope that the proposal offered to me may come to a happy conclusion. Should it thereby be justly resolved, I would wish to be able to leave here even before the beginning of the winter. In the meantime, I have also finally obtained your excellent *Theoria Motus Lunae*,[1] which I have long desired, from the local bookshop, and I have derived the greatest pleasure from reading through it. I notice that you, notwithstanding your far deeper insight in this matter, have nevertheless still come across enough difficulties in it, which is no small consolation to me for the unspeakable trouble that I have previously had with it. While my method is different from that which you develop in your treatise, it is nevertheless more in accordance with the first as with those which are contained in the appendix. Right at the beginning, I have substituted in the general equations instead of η (which denotes the true distance of the Moon from the Sun), another angle that did not depend upon so many inequalities. Hence I have selected a series of known angles with undetermined coefficients both for the true longitude of the Moon and also for its distance; and after substitution, and determining these coefficients as well as those inequalities which were not deducible with the chosen series, I have moreover discovered in particular that I actually proceeded more simply with the latter because on the whole they remained small. Nevertheless, many could not be determined with sufficient precision, the nature and cause of which imperfection is well enough known to you. I have deduced the latitude of the Moon directly from the general equations for those [inequalities] which you also quote on p. 12, without first transforming them into two others; by which [means] they have offered different advantages. If I compare this hitherto undescribed method in general with that which you use in your invaluable script, it seems to me that I have chosen a much more tedious way without having been more fortunate. Meanwhile, since I still have in order my transformed equations which have produced the most difficulty, I found the still unknown, or at any rate not precisely-enough known, inequalities implicitly embodied in them, so I had the greatest desire to extract these completely in order to improve my tables still further; only I do not see where I can acquire enough time for so doing. In addition, there is still a lack of the observations themselves, which up to now do not have the necessary degree of correctness, and thus my efforts could be in vain. I have subsequently collected many further observations, in particular from solar and lunar eclipses,[2] and they agree with my table; I am consequently now entirely assured that among those eclipses which were observed since the invention of telescopes and pendulum-clocks, there is none with which the tables do not agree to within 2 minutes in time, or $1'$ in longitude. One of the most elegant applications of this very tedious comparison is that one should now be able to predict every single future eclipse correctly to within very few seconds of time; I shall some day specify the means in a special paper.[3]

If our local transactions of the [Göttingen] R[oyal] S[ociety] have had the good fortune of coming into your hands, the instrument together with the method of using it which I proposed for accurate land-measurement,[4] will not be unknown to you. You will easily judge from it that owing to the continual twinkling of the stars it is not useful for astronomical observations. The great accuracy which can be obtained through it with so little trouble, has meanwhile given me a longing to consider such a means through which a similar small and convenient instrument could be made fit for astronomical observations. Although I now still do not know for sure whether the method which in this case I have approved will have the desired success, I have nevertheless ordered an instrument of this new design from a local artist,[5] and as soon as it is ready I will try to determine the elevation of the pole for this town in order to see how far it [i.e. the instrument] can be relied upon. I hope that one could at least observe the meridian altitudes of both the Sun and the stars to within 10″ with it, and that it consequently does as much as a common quadrant of from 3 to 4 feet in radius; although the former will be scarcely so many inches in radius.

I have the honour to remain with the greatest respect, sir,

Your obedient servant,

T. Mayer

17. *Mayer to Euler; Göttingen, 22 August 1753*

Dear Sir,
Honourable Professor

No one can be more sensitive to the delay in my desired good fortune, than myself; meanwhile my decision remains unchanged, and I will wait for the return of the president with patience.

Your view that the solar years may be unequal[1] has always appeared so fundamental to me that I would not have departed from it if I had not been compelled to do so through a more accurate investigation both of the ancient observations and of the correctness of the common chronological time-scale. It is first of all quite certain that from the time in which the observations of Hipparchus[2] and Ptolemy[3] were made, neither more nor fewer days have passed than one is normally used to count. The intercalation of a day in the motion of the Sun is not significant and one can therefore not suppose at all lightly an error in the common time-determination if one is to bring the Ptolemaic equinox into agreement with other observations, both old and new alike; however, all observations of eclipses and of the Moon certainly argue completely against such an intercalation. We have from Ptolemy not only observations of the equinoxes, but also of lunar eclipses, and one knows the number of days which have passed from a Ptolemaic equinox, e.g. from that which he claims to have observed at 2 o'clock in the afternoon of the 25 September, A.D.132, up to the next eclipse of the moon at 11.15 p.m. on 6 May of the year 133; consequently one cannot, without at the same time making the synodic months greater than they are actually considered to be, interpose a day between Ptolemy and ourselves. The general hypothesis would therefore differ from the improved one at new and full moon by about a day from the times given by Ptolemy; however, at the present epoch, both hypotheses would agree with one another. It follows further from that, that these two hypotheses must be different by half a day from one another around the times given by the Arabic astronomers, which fall approximately half-way between Ptolemy's and our epoch; that is, from the one hypothesis a given solar eclipse may have occurred about these mean times by day, from the other, however, by night. And then it is easy to perceive which of the two may be in accordance with the truth. It will namely be the one from the calculation of which the eclipse results as it then was actually observed by Albategnius[4] and several other Arabic astronomers. This is afforded, however, by the general hypothesis; on the contrary, according to the alleged improved [hypothesis] a certain visible solar eclipse, such as that which, e.g. Albategnius at Aracta and simultaneously another [astronomer] at Antioch have observed, should have been observed about 12 hours earlier, i.e. by night. No matter how one may tackle this problem it is impossible to account for a whole day or 13° in the motion of the Moon. I have tried to arrange the lunar tables in accordance with the hypothesis of the intercalation of one or two days, and to ascribe the 13° or 26° difference in the mean motion of the Moon to an acceleration; only it was not possible for me even to bring the observations from the two most recent centuries into agreement with it, without committing an error of half or even a whole hour, with eclipses which were certainly accurately observed to within a few minutes. If you want to take these arguments, which I mention only briefly, into more careful consideration, I do not doubt that you would not completely vindicate me. One degree difference in the motion of the Sun can be attributed more easily to the errors of the old observations than can 13 degrees in the motion of the Moon. Since, therefore, the time-reckoning requires no intercalary [day], one must necessarily conclude that the Ptolemaic equinox is really in error. My opinion of it is this. Ptolemy has simply based his solar tables on the equations and observations of Hipparchus. This is revealed, among other [reasons], since his previously-mentioned observations of the equinox are on the whole newer than his observations of planets. He has also brought the planetary tables into order, before he had observed his equinoxes. That could not be done, however, without first of all having had the solar tables in a correct [form]. He consequently borrowed the solar motion from Hipparchus without any particular investigation. This, however, put the length of the solar year at $356^{d}\ 5^{h}\ 55^{m}$ . . ., so large in fact, *quod probe notandum*, that it is [deduced] from its luni-solar cycle of 76 years, and not from actual observations. In my view this year is now too large by approximately $6\frac{1}{2}$ minutes, which in the 300 years which have passed by between Hipparchus and Ptolemy, amounts to approximately $1\frac{1}{4}$ day or about 1° 15′ in the longitude of the Sun. And the Ptolemaic equinoxes are found to be too late by just about as much. It can be that Ptolemy perceived this error of his solar tables in his observations of the equinoxes, which are the very last of all his remaining observations; only, because he had already built his whole system upon it, perhaps he had rather wanted to discard his observations than to attempt to revise his system from the outset. Since, however, no one could object to it, he pretended that the erroneous equinoxes of his tables were true and observed. There are more and newer examples of an astronomer, from too great a love for his constructions, falsifying observations. This is certain at least in Lansberg[5] and Riccioli.[6] How much more simply could not Ptolemy, who perhaps did not imagine that one would ever be able to disclose his deception through more accurate observations, have fallen into this error. Yet what can only be surmised from this line of evidence becomes completely certain through Ptolemy's statement about the motion of the fixed stars. From his observations of the fixed stars, which are thought to be older than [those of] his equinoxes, he concluded that the longitudes themselves may have increased by about 3 degrees from Hipparchus' to his own time, i.e. in 300 years; from that it is therefore quite certain that this slow movement amounts to 4° 13′ in the same time (for to assert an actual, and in any case very irregular variation of the equinoctial points, precludes other circumstances). Thus he pretends that the motion of the fixed stars is approximately $1\frac{1}{4}$ degrees too small, namely just as much as his tables err in the motion of the Sun. Now this must necessarily result as follows. Ptolemy observed the longitudes of the fixed stars in the old manner of that time thus: he measured by day the ☽–☉ distance from the position of the Sun, which he had to guess from a table, since at that time he had still not observed his equinox; from these and the observed distance of the ☽ from the ☉, I maintain that he determined the longitude of the Moon which consequently also came out at about $1\frac{1}{4}°$ too small. On the following night, he measured the distance between the Moon and a fixed star and thereby deduced, in the manner described, the longitude of the fixed star. Thus he also found this [to be] too small by about $1\frac{1}{4}°$. Now he compared it with Hipparchus' observation, and necessarily found no greater difference than 3°. Against this, he would have found for it $4\frac{1}{4}°$ or the true amount, if his tables or also his equinoxes which were in harmony with them had not always given [a value] for the position of the Sun which was $1\frac{1}{4}°$ too small. This circumstance is sufficient to show that no credit should be attached to Ptolemy's observations of the equinoxes.

I could quote various arguments against the secular acceleration of the motion of the planets and in particular Saturn, whose motion is not only unaccelerated but even retarded; only time and space are both used up on this occasion, and if you permit me, I will reveal my doubts next time.

Meanwhile, I have the honour to remain, sir, with eternal respect,

Your obedient servant,

T. Mayer

18. *Mayer to Euler; Göttingen, 25 November 1753*

Dear Sir,
Honourable Professor

The reasons with which you have supported the resistance of the aether are indeed so important that the small doubts which I have had against it have now been quite abolished.[1] However, so much seems to be certain, that the effect of this resistance on the Earth's motion may either be very small, or augmented by the attraction of the other planets. In the preparation of the solar tables, which are attached to my lunar tables, I have already done what you want. I have namely first of all determined, without considering the ancient observations, the length of the tropical year from quite recent observations, especially those which are recorded in M. le Monnier's *Histoire celeste* in the years 1683–85,[2] then I compared it with those which were made only a few years ago at Paris. In so doing, I have paid attention at the same time to the inequality which arises from the nutation of the Earth's axis, as its influence in such a determination is rather significant.[3] The length of the year is based only upon these observations which were introduced into the tables. Hence I reverted to the observations of Walther which were made at Nuremberg at the end of the 15th century.[4] Since one does not really know at which place in this town Walther has observed, and [since] consequently the elevation of the pole is somewhat uncertain, I therefore employed a method by which it is not necessary to know the actual elevation of the pole.[5] From these observations I again found the year to be just as long as the new observations give it; at least, the differences were not so considerable that one could ascribe them to an acceleration in the motion of the Earth, particularly because [not only] did these sometimes show an excess but sometimes [also] a deficiency, and are consequently to be attributed rather to the incorrectness of the observations. I investigated further the observations of Albategnius[6] and finally those of Hipparchus. All of these agreed very well with the previously-established length of the year, as would not have occurred if I had assumed an acceleration. [The fact] that Cassini, in his *Elémens d'Astronomie*,[7] found several discrepancies in the length of the tropical year, which seems to favour an acceleration, I would consider to be a rough approximation, and his method is also not precise enough. After such an accurate and varied investigation I trust myself to assert from my solar tables that they also agree as precisely with the old tables as these observations do among themselves.

However, what I have particularly noticed hereby is this, that it seems as if the eccentricity of the Earth's orbit has previously been much larger than it is now. Not only Hipparchus' equinoxes, for which the difference can easily be ascribed to another and immediate cause, but especially the old observations of lunar eclipses show traces of this, as the latter are in better agreement with the lunar tables if one assumes a larger eccentricity. Should this conjecture, which still demands more detailed investigation, become established, it would perhaps cause the motion of the Earth to be retarded through the attraction of Jupiter or also of Venus, just as such a retardation in Saturn is combined with a diminution of its eccentricity. And since the observations show no actual retardation for the Earth, one must therefore conclude that the resistance of the aether again compensates such [an effect].

The very marked acceleration in the Moon's motion may now also be best accounted for by this resistance. Yet I think that the cause might also lie in the attraction of the Sun, from which the rest of the inequalities in the Moon's motion originate. For just that reason, some of these inequalities become appreciable, because their periods are very long; thus it could happen that in the combination of so many angles, if the approximation would be carried very far, an inequality would be found whose period may be very long and [even] infinitely large. And in these instances, this inequality would transform itself into a continuous acceleration. Meanwhile, it appears to be very difficult or quite impossible to derive this [conclusion] from the theory itself.

As important as the question is, whether no inequalities are to be found in the diurnal motion of the Earth, it has always seemed to me as difficult to investigate. The means by which I have reached the decision itself, are permissible to apply only in the case when the inequalities are not all too small; and they appear to me to testify to an invariable velocity, or rather, they show how large the inequalities, should they appear at all, could not be; on the other hand, a very weak inequality can actually be in existence without one being able to detect it in this way. Since I am therefore only in the position of giving a very incomplete solution of the question, I doubt very much whether it is worthy of being presented to the Royal Academy [in Berlin].

The review of Adami's prize essay[8] in our local learned journals did not originate from me, as I have also not hitherto had any contribution to these papers at all. As far as I am able to find out, Prof. K. in Leipzig[9] has written it. A short time ago, I was instructed to collaborate on these journals, and at the same time upon the local relations regarding new books, in which I have uptil now prepared nothing other than the article on the Peutingerian table.[10]

Could I not, through your very kind help, get hold of M. le Monnier's *Observations de la Lune*, which were published in Paris in 1751? The wasted effort which I have applied to obtain that from the local booksellers, necessitates both my taking this liberty and my request. I hope to be able to derive a great deal out of it for the further improvement of my tables.

Your *Theoria* [*motuum*] *Lunae* was also recently transmitted to me from Leipzig, for which invaluable present the Royal Academy as much as I myself in particular returns its best thanks to you, of whom I have the honour to be, with the greatest respect,

Your obedient servant,

T. Mayer

19. *Euler to Mayer; Berlin, 26 February 1754*

The time of the return of our President is now gradually approaching, when he will then take the trouble of working out with His Majesty such an allowance of a Pension, that by this means the acquisition be assured for our academy of such a member, from which this [academy] itself can be promised so many advantages and fame. Which I shall not fail, in the continuation of my promises, to do everything possible to promote.

In the meantime, I have procured for myself the *Acta* of the Göttingen Society, and read in them with the greatest advantage and pleasure your dissertations[1] which bear the most pleasurable witness to all your deep insight both in the theory and practice of astronomy and double my desire of procuring such a worthy man for our academy.

Your *Tabulae Lunares* cannot be otherwise regarded as the most admirable masterpiece in theoretical astronomy, and I should never have guessed that by this means the tables would have been brought to such a degree of completeness. As I have been so little fortunate in this task, I therefore value it all the more highly, because I first of all placed the lunar tables on the footing that they were plagued before everything else by the variable eccentricity and the astonishing motion of the apogee, which things would have been impossible to harmonise with the true theory. Now you have done everything that from a practical point of view can ever be desired; only in the abstract theory do I still see much incompleteness. You will certainly also have noticed that if the Moon were 4 or 5 or more times farther away from us than it actually is, if its eccentricity were several times larger, and if its inclination to the ecliptic were also far greater than it actually is; then, I maintain, its motion could no longer be determined through such tables. I would be very curious to see tables for such a Moon, and in spite of most people considering such a work to be superfluous and almost laughable, out of such a theory the greatest advantages could nevertheless be drawn. I must confess that I do not at all foresee the possibility of the same [kind of] tables being completed, through which one could also determine only in the crudest way the position of such a Moon. Perhaps such an investigation quite exceeds our energies, and perhaps this is the cause of there being actually no such body in existence in the world, whose motion would always remain quite unknown to us. Hence one can infer a particular reason for wondering at the wisdom of the Creator, for it could have perhaps happened otherwise that such a heavenly body were actually in existence, that one could regard in all respects neither as a primary planet nor as a satellite. The motion of the Moon and of Saturn therefore appear to determine the limits of human understanding, and I do not know whether I ought not to attribute to Saturn a higher degree of difficulty than to the Moon; at least, the longer I consider this [problem] the more unsurmountable it becomes, and I very much doubt whether one will ever carry it so far that he could determine its position accurately to within a minute; so much is certain, that this can be expected from no one other than yourself.

That you have also by means of the Sun found my lunar equation to be valid has also pleased me not a little, and regarding the resistance of the aether (which still appears to me well established and necessary) I would be rather dubious since I must confess that the discovered accelerated motion of the mean Moon could easily be the effect of an inequality whose argument may have a period of many centuries. With regard to the Sun, I am completely convinced of my error; it is nevertheless to be regretted that one has no older and more accurate observations. I am taking the liberty of communicating to you a letter from China which was itself sent to me from England,[2] from which it is to be seen how the Jesuits have wanted to correct the ancient epochs through their lunar tables. Notwithstanding, it is now to be wished that they had reported the old observations properly without their corrections, so that you could perhaps still have drawn useful information from them. Since the letter to Professor Segner hardly increases the packet significantly, you will kindly excuse it [being put in the same packet] and forward it properly.

I have the honour with the greatest respect to be,

L. Euler

20. *Mayer to Euler; Göttingen, 6 March 1754*

Dear Sir,
Honourable Professor

I am very obliged to you for the undeserved attentiveness towards my moving to Berlin. The desire to enjoy your personal acquaintance becomes livelier as the time draws near when this matter should reach its end, and I have already often wished to agree to poorer conditions, so that this change should be liable to fewer difficulties. Meanwhile, I flatter myself with the hope that your recommendation and introductions to President de Maupertuis will give to this effort the effect which I desire. For my part I shall accept everything which is possible without being too much to my disadvantage.

Since I have had the honour of addressing my last letter to you, I have applied most of my spare time to improving the theory of the Moon still further, and I so little regret the many efforts that I have made so far, that I rather thought of pursuing this work still further. The fortunate success which I have already had out of it, allows me to hope for still more. Through the correction that I have just found from the observations, my tables now already agree with the truth more accurately than to within 30″. This might perhaps appear to you to be unbelievable, since still larger errors are generally involved in the observations themselves; and you will certainly be able to judge from the structure of the method which I followed that I do not promise too much. Since it does not seem possible to determine the Moon's position more accurately than to 1 minute through the observations, I have discarded this kind of observation and chosen none other than that which gives the position of the Moon through the occultations of the fixed stars and through solar eclipses. The great correctness with which the instants of such a phenomenon can be noted, is known to you. In order, however, that the uncertainty in the place of the fixed star might have no appreciable influence on the position of the Moon, I have first of all collected only the occultations of a single fixed star; then also if there were already a few minutes of doubt in the position of this fixed star, the error would certainly not prejudice the use of these observations in the investigation of the inequalities of the Moon's motion, but merely be reflected in the epoch of the mean motion of the Moon, which one can afterwards easily correct through the observations of the solar eclipse together with the position of the fixed star. We have a considerable number of observed occultations of Aldebaran; out of these I have calculated the positions of the Moon with the help of the parallax, so that there can be no error of 5″ or 10″ in them. After this, since I had hereby found the greatest corrections, I have similarly made use of the occultations of other bright stars such as Spica, Regulus, Antares, the Pleiades, etc., and already deduced from them over 100 such very precise positions of the Moon, which have given me still more, although almost inappreciably small, corrections of those equations which I employ in my printed tables. Since all these positions of the Moon are exceptionally accurate, and the tables already possess an internal accuracy of under 30″, they give a much more decisive confirmation of the truth of these corrected tables, than the general observations in which I have often, to my great annoyance, detected errors of several minutes. If I had only more uninterrupted time to devote to this, which I hope to receive in Berlin sooner than [I would] here, I should not regard it as difficult to reduce even these small errors to about half [of their present values], and thereby to give to the theory of the Moon, the highest degree of accuracy of which it will ever be capable.

I have just been able, with much less effort, to bring the latitude of the Moon to just this degree of accuracy. And this [was] in fact [done] not so much through the observations, as rather through the theoretical calculation and further development of the method followed in my printed tables. This method, as I have already stated in one of my earlier letters to you, consists in a particular solution of the IIIrd equation which you have given for the Moon's motion, and which is to be found on p. 10 of your *Theoria* [*motuum*] *Lunae*, to which I shall refer for the sake of brevity. I do not divide this equation into two others, but solve it directly; and the manner of so doing is the following of which, however, I illustrate the principal part, in order not to exceed the limits of this letter. If, in the IIIrd equation p. 10, $x \tan \psi$ is substituted for r, this equation, still without expanding it, has this form: $\mathrm{dd}(x \tan \psi)\frac{1}{2}R\, \mathrm{d}t^2$. Since ψ signifies the true latitude of the Moon, the most important term in this same expression must nevertheless have this form, $A \sin \alpha\varphi$; where A denotes the tangent of the mean inclination, φ the longitude of the true ☽, but $\alpha - 1$ the mean motion of ☊; thus I first put

$\tan\psi$ [=] $A\sin\alpha\varphi + v$; then, and through the use of the first two equations page 12 *loc. cit.*, equation III is transformed into:

$$O = \mathrm{dd}(vx) - (\alpha^2 - 1)Ax\,\mathrm{d}\varphi^2 \sin\alpha\varphi \\ \qquad - \tfrac{1}{2}\,\mathrm{d}t^2 AV \sin\alpha\varphi - \tfrac{1}{2}\,\mathrm{d}^2 t\, AT\alpha\cos\alpha\varphi + R$$

in which most of the symbols have the meaning which you assigned to them.

If in this equation the appropriate approximate values with the exclusion of the terms which depend upon the Moon's parallax, are substituted for dt, V, T and R, and if one further assumes for v this series of terms: $+B\sin\beta\varphi + C\sin\gamma\varphi + D\sin\delta\varphi$ + etc., the following equation is deduced:

$$O = \begin{cases} +A\left[1 - \frac{2}{3}(\alpha^2 - 1)\dfrac{y^3\,\mathrm{d}\varphi^2}{k^2\,\mathrm{d}r^2}\right]\sin\alpha\varphi - \\ -A\left[\dfrac{\alpha+1}{2}\sin(2\omega - \alpha\varphi) + \dfrac{\alpha-1}{2}\sin(2\omega - \alpha\varphi)\right], \\ +B\left[1 - \frac{2}{3}(\beta^2 - 1)\dfrac{y^3\,\mathrm{d}\varphi^2}{k^2\,\mathrm{d}r^2}\right]\sin\beta\varphi - \\ -B\left[\dfrac{\beta+1}{2}\sin(2\omega - \beta\varphi) + \dfrac{\beta-1}{2}\sin(2\omega + \beta\varphi)\right], \\ +C\left[1 - \frac{2}{3}(\gamma^2 - 1)\dfrac{y^3\,\mathrm{d}\varphi^2}{k^2\,\mathrm{d}r^2}\right]\sin\gamma\varphi - \\ -C\left[\dfrac{\gamma+1}{2}\sin(2\omega - \gamma\varphi) + \dfrac{\gamma-1}{2}\sin(2\omega - \gamma\varphi)\right], \\ +D\left[1 - \frac{2}{3}(\delta^2 - 1)\dfrac{y^3\,\mathrm{d}\varphi^2}{k^2\,\mathrm{d}r^2}\right]\sin\delta\varphi - \\ -D\left[\dfrac{\delta+1}{2}\sin(2\omega - \delta\varphi) + \dfrac{\delta-1}{2}\sin(2\omega + \delta\varphi)\right] \end{cases}$$

etc. etc.

In this, $k:1$ = mean motion of the ⊙: mean motion of the ☽
$\mathrm{d}r:\mathrm{d}\varphi$ = mean motion of the ☽: true motion of the ☽
$1:\alpha - 1$ = mean motion of the ☽: mean motion of ☊
$1:y$ = mean distance of the ⊙ from the ♁: true distance of the ⊙ from the ♁
ω = Longitude of the true ☽ − longitude of the true ⊙
A = tangent of the mean inclination of the ☽s orbit to the ecliptic, whose value is derived from observations

The values of β, γ, δ, etc., are to be chosen according to the circumstances;

they are easily obtained. α, B, C, D, etc., however, are determined if one sets all the terms $\sin \alpha\varphi$, $\sin \beta\varphi$, $\sin \gamma\varphi$, $\sin \delta\varphi$, etc. $= 0$. The expansion of this equation, which, if it is multiplied by $\frac{3k^2\, dr^2}{2y^3\, d\varphi^2}$, possesses no particular difficulties, first gives me a small correction to the mean position of ☊ of this form $+9'\,6''\sin s - 7''\sin 2s$, where s represents the mean anomaly of the ⊙. If the position of ☊ is found through this correction, the true distance of the ☽ from this ☊ $= \delta$, and the anomaly of the mean ☽ is put $= p$, then:

The northerly latitude of the true ☽

$$\begin{aligned} &= \text{the latitude of the ☽ dependent simply and solely upon the angle} \\ &\qquad \delta + 8'.46''.6 \sin(2\omega - \delta) \\ &\quad - 2''.1 \sin(2\omega + p - \delta) + 24''.7 \sin(p - \delta) \\ &\quad - 9''.7 \sin(2\omega - \delta + s) + 16''.5 \sin(2\omega - p - \delta) \\ &\quad + 23''.0 \sin(2p - \delta) - 3''.6 \sin(2\omega - \delta - s) \\ &\quad - 6.4 \sin(2\omega - 2p - \delta) - 2.7 \sin(3p - \delta) \\ &\quad + 2.1 \sin(2\omega - 2p + \delta) \end{aligned}$$

in which no term amounting to more than 1″ is omitted. The observations confirm the correctness of this formula so well that I am still unable to find any difference of more than about 10″.

Since I, in the manner described above, can determine rather accurately with the help of my tables the true place of the star from the occultations [of the same], I have by this means discovered different errors in the catalogue of fixed stars which are so considerable that it is no wonder that astronomers are still in error by about several minutes in the position of the Moon, which is indeed much more difficult to observe. The catalogue which H. Zanotti in Bologna prepared a short time ago from the fixed zodiacal stars and has published with the ephemerides from the years 1750–61[1]—and which appears to be made with a great deal of hard work and with the most accurate instruments—is indeed rather accurate in [the case of] the northern stars. On the contrary, with the stars which lie in the southerly part of the zodiacal belt, he has generally about a whole minute too much. The longitudes of the fixed stars, which M. le Monnier quotes in his *Histoire céleste* [Paris, 1741] and again in his *Institutions astronomiques* [Paris, 1746] from his own observations, are similarly not always correct, and [this] has led him to commit the error in the position of Regulus, of attributing to this fixed star a slow retrograde proper motion which, however, I do not find [to be] in agreement with the truth.

So that you can make a judgment on this matter, I will compare the longitudes of 3 of the most conspicuous fixed stars of the zodiacal circle which I have found from my own researches, with those which Messrs Zanotti and le Monnier have estimated for them.

Beginning of 1750					Zanotti				Le Monnier			
	s	°	′	″	s	°	′	″	s	°	′	″
Longitude of Aldebaran	2	6	17	55	2	6	17	51	2	6	17	20
Longitude of Regulus	4	26	21	41	4	26	21	44	4	26	20	45
Longitude of Antares	8	6	16	35	8	6	17	29	8	6	17	20

Since up until now one has regarded as the best way of observing the position of the Moon, the method of comparing its meridian transit with the transit of a star positioned not far from it and then deducing the Moon's position from the position of the fixed star, the errors which still exist in the position of the fixed star must have necessarily had a full influence on the observed position of the Moon, and it is consequently easy to guess how little one can place one's trust in such observations for improving the theory of the Moon, if they were not published in detail together with all the [relevant] circumstances.

I regard just this as one of the principal obstacles why one still does not know very accurately the theories of the planets, particularly of Saturn and Jupiter; and supposing that one had also completely exhausted the theoretical calculation for Saturn and Jupiter, one would certainly never be able to prove its correctness, so long as one has still to worry about such appreciable errors in the observations. I am entirely of the [same] opinion as you concerning the theory of Saturn, and regard the determination of its motion [as] much more difficult than [that] of the Moon. It would, however, be still more difficult, if you had not already cleared the greatest obstacle out of the way, and at the same time broken the ice, in your admirable prize essay.[2]

The consideration of such a moon, which was still further away from the Earth than our Moon, and which had a greater inclination to the ecliptic and a greater eccentricity, would certainly bring much advantage to theoretical astronomy. The opportunity could also certainly occur if a comet of the Earth or of another planet came very close. The investigation of such a motion would, however, be an everlasting cross for astronomers [to bear].

I return my most sincere thanks for the communication of the letter of P. Gaubil;[3] as soon as I have studied it with proper attention, I shall send it back to you. Meanwhile, I can state so much, that according to my tables the solar eclipse of the year 2128 B.C. was invisible in the whole of China.

The Chinese reports seem to me very suspect, or at least to require a severe criticism.

I have the honour to remain with the greatest respect,

Your most obedient servant,

T. Mayer

21. *Euler to Mayer; Berlin, 11 June 1754*

Any day now we expect the return of our President, which should quite definitely occur before the end of this month. He will at this very moment be doing his utmost to obtain a proper salary for you worked out with His Royal Majesty; and I very much hope that this may have already been done and that M. de Maupertuis will not have considered it necessary to delay his return on account of the death of his brother. This delay is all the more difficult for me since I hold a formal Order as a member of the Russian royal academy from its illustrious president, to offer you an appointment in it.[1] The astronomical positions are actually already filled, one merely offers you the choice from the vacant places which remain; these are first of all my previous position Higher Mathematics, [and] after that, Physics and Mechanics, in which I have still to appoint a director for the Geography Department. You have already shown satisfactorily that you are equally suitable for this post, and if in fact you are keen to choose one of them I ought to assure you that in consideration of your particular earnings no difficulty will be made in fixing the salary to exceed the usual [amount]. You can reckon on at least 1000 rubles yearly, and the cost of travel should also be richly paid in advance. You could be engaged for 4 or 5 years and after this time either stipulate a resignation or an increase of salary, the latter of which could also come about after 2 or 3 years. You ought to have much less doubt regarding the agreeableness of the way of life there, since the academy is now quite particularly linked to Her Royal Majesty,[2] and also to date no one from the academy has had cause to complain. This is now what I ought to have told you by order of Mr Hetman[3] and Excellency the Lord President, and on which I must elicit a favourable reply.

I would like to wish that our local academy were in similar circumstances and could promise you just as advantageous conditions; however, perhaps this very circumstance that would attract you to Petersburg can give the matter greater weight. Thus I wanted to advise you respectfully not to come to a final resolution immediately, but to postpone the matter for at least a couple of months, in which time all useful efforts can be applied here (which may otherwise not occur); it could then stretch to 6 weeks before I receive your answer, send it to Petersburg, and receive new instructions in return.

In coming now to your new important discoveries, I first of all congratulate you wholeheartedly upon them, and wish that their importance would soon be known to everyone. Your first tables[4] have indeed aroused as much applause as amazement, only jealousy has already let itself be seen more than clearly. You have undoubtedly already seen M. d'Alembert's new treatise on the Moon,[5] in which he refers in a contemptuous way to the tables and mocks the previously given exact determination itself; he does not believe that you have made use of the theory and that he and M. Clairaut were the first who investigated the theory of the Moon. Notwithstanding my already long published tables whose quite new construction permits the theory to be easily satisfied, although the errors are still too large, he will deny me almost all parts [of it]. You would not be behaving badly, if you by chance refuted this braggard to the fame of the Germans. In England one seems to have been fairer since, following the representations which I made on the importance of your tables, I have received such an answer from which I can with reason conclude that you would be regarded as a worthy competitor for the award established for the discovery of longitude. If you could now still add a method of knowing how to determine at sea the position of the Moon through its distance from fixed stars so accurately that the complete longitude [obtained] from the comparison of it with your tables does not vary by more than half a degree from the truth, you could be assured of the prize of 20,000 pounds sterling.[6] It would, however, also still be necessary to determine the true positions of more fixed stars more accurately. Then, since you brought things so far through your untiring industry that your tables denote the longitude of the Moon accurately to 30″ and the latitude even to 10″, there is no further improvement to hope for from this side: and Mr Dalembert himself stated that his tables are still over 5′ in error.

22. *Mayer to Euler; Göttingen, 23 June 1754*

Dear Sir,
Most highly esteemed Professor

To your most kind offer, concerning a post at the Royal Academy at Petersburg, I am unable at present to give a decisive reply. I shall certainly follow your well-meant advice and postpone my final decision, if it is otherwise permissible, until such time as I see how things will fare in Berlin as far as I am concerned. Nevertheless, I have the honour to assure [you] that I would prefer Berlin under very much poorer conditions, if only they are provided so that I can assume them without disadvantage. And since this is my true disposition, you yourself will know best how the answer to Petersburg may be framed, with which the postponement of a positive answer could best be excused. It would actually not be difficult to persuade me to give the proposals from Petersburg a hearing if there were no condition with Berlin and in particular the pleasant hopes of likewise profiting from your closer acquaintance.

I have indeed already learned a few months ago from Herr von Haller,[1] and also still more recently from Prof. Stähelin[2] from Basle who honoured me with his visit on his journey through Göttingen, that Mr d'Alembert found fault with my lunar tables. I am, however, so unfortunate that despite searching everywhere, I have been unable uptil now to get hold of the book in which this attack occurs, and I also see no hope of being able to obtain it. Should you [be willing to] dispense with your copy for a couple of weeks, I would be so bold as to beg you most humbly to communicate and mail it here, the postal charges of which I would bear with the utmost willingness. If M. d'Alembert has nothing else against me than that of which you have had the kindness to inform me, I shall certainly not lose anything through this affair. In order that my printed tables always attain [an accuracy of] up to 1′, I have furnished evidence with such an enormous amount of observations, in volume III of the *Comm. S.R. Gött.*[3] which has recently been published, that a still greater disbeliever than M. d'Alembert can be convinced by it. And that I have taken the theory to [my] assistance, even before I have seen anything at all about it either from M. Clairaut or from M. d'Alembert, except simply that which the former wrote about it in the 1743 or 44 *Mémoires* and which is certainly bad,[4] I can prove partly through a quantity of papers lying beside me, partly through my friends in Nuremberg, partly through letters to M. de l'Isle which I wrote to him in the years 1749 and 50,[5] and partly through what I have told you about it from time to time. When the lunar theory was proposed for study to the scholars of the Petersburg Academy,[6] I wanted to send a solution there; however my locally-held vocation did not leave me with enough spare time to bring this into such an order that I could present it to the Academy. Apart from the theory, I have also freely employed a vast quantity of observations, as much in order to determine more accurately the elements themselves, as [to determine] several equations which the theory leaves in doubt; but I do not see how such blame could be placed upon me. Mostly, however, I wonder how M. d'Alembert can be so unfair towards you, since the whole world must certainly perceive and acknowledge that M. Clairaut as much as all others who did something about this for the past ten years, have made use of your important and fortunate discoveries and skilful application of the analysis on the lunar theory, and that without your researches in this aspect of astronomy the true theory would perhaps still lie hidden. I throw myself most unwillingly into conflict, since it always costs more time than it is worth, yet I certainly regard [it] as necessary this time to defend myself openly and to refute the unfounded and jealous allegations of M. d'Alembert; and I shall do it, as soon as his writings become known to me.

I am indeed assured that my tables, as I have now improved them, are more than sufficient to find the longitude at sea to within $\frac{1}{2}$ a degree and still nearer. I have also already determined very accurately the positions of all stars of the 1st, 2nd and 3rd magnitudes which lie in the zodiacal circle,[7] and found in addition a simple method of calculating its [i.e. the Moon's] true position correctly by a single distance of the Moon from a star.[8] The only major question is, whether the sea-octant can be used on the ship so that one can observe the distance of the Moon from a fixed star accurately enough, namely to within 1 minute. The correctness of these observations depends not only upon the construction of the instrument, but also on the rocking of the ship. With the first, it is possible to estimate the error; however, the other factor is so variable that one can make no general prediction of the error that can arise from it. In a calm sea the observations are undoubtedly made more surely and accurately than if it is rough. And since I therefore cannot prove except by probable reasons, that this method of finding the longitude through the help of the lunar tables is sufficiently correct, I have not previously trusted myself to write anything about it to England, still much less given myself some hope of obtaining the reward for it.

I have the honour to remain with the greatest respect,

Your most obedient servant,

T. Mayer

23. *Mayer to Euler; Göttingen, 17 August 1754*

Dear Sir,
Honourable Professor

In considering your last proposals, I have taken rather a long time for reflection, and to tell you the truth I am presently still largely undecided. I considered the circumstances on the one hand, and I find that I cannot accept and exchange against my local position the proferred salary of 600 Rthl. without a marked loss. Ever since several changes [which are] advantageous for me have occurred in our local scientific society, my normal income has been increased so much that I am still inclined to hesitate before taking up another post under 700 Rthl. However, on the other hand the particular inclination towards the academy [in Berlin] together with the pleasure and advantages which I am promised from the contact with so many scholars and famous people, as well as the peace in which I hope to be able to apply myself to science there, also makes me willing to resign my local post with several disadvantages. Through such circumstances, which seem to keep me on both sides of the scales, I do not know what to decide. Meanwhile, if I should make a final declaration right away, it would stand out as follows. You, I hope, can take away to your satisfaction, how willingly I wanted, in view of the above circumstances, to do the utmost that is possible for me. Namely, I would without further consideration accept the position offered to me by the academy there if a yearly salary of 650 Rthl. were allocated to me; and in addition to this, if the spending of the travelling expenses of 100 Rthl., which I believe will scarcely be sufficient, will be presented immediately after my arrival in Berlin. As to my resignation, I see no difficulty in obtaining this since it would largely depend upon myself. I would not, however, be able to leave before October, since first of all the College closes in September, and I need some time to prepare myself for the journey, to sell my spare furniture which is inconvenient to transport, and so on. On the other hand, I would not demand that the start of my pension begins earlier than my departure. Thus you see here my final sincere resolution, and if these conditions made as convenient as possible by me should be rejected, I have therefore lost all hope of coming to Berlin, with which I have hitherto flattered myself so much.

The lack of time prohibits me on this occasion from informing you, as I originally intended, about the good progress of the researches which I have made up to now on the Moon's motion. Hence I close and remain with the greatest respect,

Your most obedient servant,

T. Mayer

24. *Euler to Mayer; Berlin, 27 August 1754*

Honourable Sir,
Most highly esteemed Professor

Our President has considered your last explanation [to be] so fair and most highly modest that he not only granted the proposed conditions by express order of His Royal Majesty, but the yearly salary has been fixed at 700 Rthl., apart from which you are to receive not only the desired 100 Rthl. for the journey but also free accommodation afterwards, with the definite assurance that as soon as the circumstances permit, you are to receive an increase in salary. From this you can see that one knows here how to value your extraordinary merits; and the same will not be lacking in future to regard [you] more according to your value. On the basis of this you can therefore go ahead with all arrangements on the maintenance of your resignation and departure, so that you will certainly arrive here in the coming October and can take possession of your post. Our observatory ought then to be completely handed over to you, so that you arrange the observations in accordance with your own intentions, apart from which nothing further will be expected from you than that you attend the weekly meeting of our academy and from time to time lecture on your discoveries and works. Mr Kiess[1] has also up to now published an astronomical calendar annually; since he now received a call to Tübingen to the post of the late Prof. Krafft,[2] however, it will depend upon you whether you want to continue this work. I therefore congratulate you wholeheartedly on this happy end to our previous correspondence and hope soon to have the already long desired pleasure of knowing you more intimately, and to enjoy actually sharing your friendship. I shall contribute with the greatest pleasure everything which will be in my power to make your stay here enjoyable, in particular however to simplify the initial installation of [one to] whom I have the honour to be with the greatest respect,

Your lordship's most obedient servant,

L. Euler

Kindly try to transmit for me the enclosure to Professor Segner, which contains nothing about our business.

The enclosed letter from the President himself under the academic seal gives you a full guarantee about everything.[3]

P.S. Even if Mr Lowitz in Nuremberg should not resolve to accept a call to St Petersburg,[4] I shall no longer make any more propositions to you; since I am assured, that you in future will never acquire any desire to leave Berlin.

25. *Euler to Mayer; Berlin, 31 August 1754*

Honourable Sir,
Most highly esteemed Professor

It is fortunate for our local academy that the Royal Academy of Russia has delayed so long with its proposals to you, and I hope that you yourself will not consider the actual circumstances of this delay otherwise. You could indeed cut a dash on a much more substantial salary, especially since it also increases much more; but after the passage of several years you would certainly receive a desire to return again to Germany, and there it might not be so easy to find a decent post. I hope at least that you will be completely content with the most recently transmitted conditions, and since in the meantime our Mr Kiess[1] has actually tendered his resignation, you could hope all the quicker for an improvement, and be assured that your hard work and zeal for the improvement of our academy will be honoured. You will surely give to the enclosed letter an answer which will not be disadvantageous to the academy here; and also explain it to Professor Heinsius[2] in such a way that we here can be assured of seeing you with us in the coming October, in which pleasant hope I respectfully recommend myself to your treasured friendship and remain with the greatest respect,

Your most humble servant,

L. Euler

26. *Mayer to Euler; Göttingen, 7 September 1754*

Dear Sir,
Honourable Professor

I return to you my most sincere thanks for the extraordinary good fortune obtained for me in Berlin, and hoped to be in a position to be able on this occasion to give proof of my obligation in a positive way. I hoped to be able to give you today the news of my resignation, for which I have immediately applied;[1] but in order that no difficulties might actually be made about it, it will still be subject to [what is] for me an annoying delay. As I now understand, our most gracious King's own consent is required for it, and consequently [it has to] be sent to England, which will take at least three weeks. Meanwhile, I will make all the preparation, so that I can leave immediately after the received resignation. I hope in Berlin to find the original observations of the late Mr Kirch[2] which, as I suppose, could at least be made known in print, and to find among these various ones [that would be] useful to me. I would willingly take the preparation of the calendar upon myself and give it a construction which could make it still more popular and more convenient. I would, in particular, introduce into it the Moon's motion according to my new tables,[3] which are now certainly not more than about half a minute in error, as I can substantiate by several hundred selected reliable observations. One of my first duties in Berlin will be to transmit these tables to the press, together with an adequate explanation and justification. Since the [last] time [I wrote], a marked improvement of Hadley's sea octant has occurred to me; and I am actually having an instrument with this improved construction made for me,[4] in order to see whether it yields the necessary accuracy at sea for the distance of the Moon from a star. The method of finding the longitude from a single distance of the Moon from a fixed star (which must, however, be in the zodiac) requires that one first assumes the longitude as known and then calculates from the tables the distance of the Moon from a star, after which he compares this with the observed [distance]: and if it is not in agreement with it, the true longitude is derived from it through the aid of a new supposition according to the *regula falsi*.[5] In this way, a great deal of effort and uncertainty is avoided. The good fortune of having in Berlin a completely equipped observatory at my complete disposal makes me want to forget the local observatory all the earlier, since due to the lack of a competent instrument-maker, the instruments for the latter will still require much time.

Whether Mr Lowitz in Nuremberg will accept a call to St Petersburg, I cannot say. The large globes whose manufacture he took upon himself, would certainly be the greatest hindrance in keeping him back from it.[6] Otherwise, I wish him such a good fortune, which he certainly deserves.

In the pleasant hope of soon being able to testify verbally to my most humble gratitude for the fortune which I have realised, I have the honour to be with the greatest respect,

Your most obedient servant,

T. Mayer

27. *Mayer to Euler; Göttingen, 15 September 1754*

Dear Sir,
Honourable Director

The enclosed letter is an answer to the offer from the Royal Russian Academy which I most humbly request you to transmit.[1] I have had no hesitation in rejecting this offer (which I nevertheless treasure as a special honour) since I am so completely satisfied with my new appointment in Berlin, and desire no more than to receive soon my resignation with which I am still no further advanced since my last [communication]—in order still to be able to make my journey there before the bad winter weather. As soon as I receive some definite news regarding this matter, I will not neglect to inform you of it, so that a formal call can be sent out to me to my complete surety. Meanwhile I have the honour to remain with the greatest respect,

Your most obedient servant,

T. Mayer

28. *Mayer to Euler; Göttingen, 6 October 1754*

Dear Sir,
Honourable Director

If anything has been rather difficult for me, it is the writing of this letter, in which I must inform you that I am not coming to Berlin. I have applied for my resignation in the most positive manner possible, but no one wanted to be given any explanation of it until I myself gave the conditions under which I wish to remain here. You can easily judge that I have not made these weak, and they have been complied with against all my suppositions, so much so that even the directorship of the local observatory, which was previously shared, was given over to me alone.[1] Circumstances have been involved in this matter which, as it appears, have perhaps caused me to be thought of more highly than I deserve. And fortune has not wanted me to be able to take advantage of the benefits which you had planned for me in Berlin, and for which I shall nonetheless be eternally most indebted to you. In this connection I am most grieved that the hope of my coming into closer contact with you has come to nothing; I would be much more grieved still, however, if you would now become impatient with me and deprive me of a part of your invaluable favour and written association.

Perhaps at another time it will be possible to bring about that which now, due to insurmountable obstacles concerning my position can not be realised. And should I moreover always be deprived of this good fortune, I shall certainly never cease to dedicate to the academy which has honoured me with the status of a member of it, [and] in particular to yourself, the most thankful and most respectful loyalties, in which I remain with the greatest reverence and respect,

Your obedient servant,

T. Mayer

29. *Mayer to Euler; Göttingen, 1 December 1754*

Dear Sir,
Honourable Director

I have been obliged on account of all kinds of obstacles to leave your letter of the 22 October unanswered until now, and before everything request that you do not interpret this delay as a sign of a hostile affection that this might have produced in me. On the contrary, I recognise that you could have cause to be displeased with my conduct in the former vocation matter, as there is nothing more unfortunate in it for me than that it caused you some annoyance—[a circumstance] which I am not, however, in a position to brush aside—and thus can do nothing more than to beg that you [do not] credit me with this false step. It is indeed true, that I have explained myself concerning the taking up of the post proposed to me [in Berlin] without much reservation or subterfuge, as one does if one has firmly decided not to allow himself to be held back in any way. Since I, however, certainly still had no formal call in my possession, and I could easily perceive that this would be held back out of some mistrust, I therefore also thought of always giving myself so much room for manœuvre that I could retract in any case, without blaming myself for wanting to deceive anyone. If it were necessary to produce witnesses for my relations in this matter, these would quite clearly show how little one could blame me in this for some lies or the like. The way in which I sought my resignation, was to my mind so formal that another [person] without a proper vocation in his hands, would perhaps not have taken such a risk. I suppressed this circumstance, however, so that one could have no reason for advising me against it. Certainly, as you yourself concede, I have never had the intention of going behind anyone's back, so I will not excuse myself further. If I have already done so inadvertently, it is rather for want of sufficient resolution, than for any other reason. The circumstances under which I remain here are, as I can indeed believe, a secret, and if you have been informed that I have been deceitful, such has been written contrary to the truth, perhaps from someone who believes himself to be hurt through my remaining here.[1]

Since you want to honour [me] still further with your priceless good will, although you could have cause to deprive me of it, I am therefore writing this to your known affability, and thus make use of the very kind permission [that you have] granted, to continue our previous correspondence.

For the repeated offer which you are pleased to make to me from the Royal Russian Academy I am very obliged, but it is a pity for me that I cannot make up my mind to accept it.[2] Concerning the maps and other matters from Russia, I have written to Nuremberg and am still awaiting the reply.[3] As far as I know the Beauplan maps of the Ukraine are available there; however, they are regarded as a rarity and they would certainly be given away reluctantly since only one copy is available.[4]

I have recently read with great pleasure in the Memoirs of the Berlin Academy your explanation of colours which originate through the gratings on glass plates;[5] and wondered at the agreement of your theory of light with this phenomenon; although I was already convinced beforehand of the truth of this theory. I believe [that] the objection made by someone against this theory was whether it could completely explain how it comes about that the light rays are cancelled if they go through an opening—not bent to the side, as in the case of sound—due to the so-called modification of the rays; which modification is actually nothing else than a small deviation of the rays from the straight line, or according to your theory an expansion of the pulse, if it must pass close to a dark body. The ring which one sees around the moon in the total solar eclipse, the coloured stripes which one observes around the shadows of every dark body, and all similar appearances are, in my opinion, easy to explain by such a broadening and expansion of the pulses. Perhaps with time we have to expect from you a detailed explanation of this phenomenon.

I have the honour to be with the greatest respect imaginable,

Your most obedient servant,

T. Mayer

30. *Mayer to Euler; Göttingen, 23 February 1755*

Dear Sir,
Honourable Director

What you have had the kindness to inform me about, concerning Mr Bradley[1] (that my lunar tables are highly regarded by him) has been all the more pleasant and understandable, since I have at just about the same time sent the manuscript of these improved tables to England;[2] where it will now undoubtedly come into the hands of Mr Bradley himself, as much for [the purpose of] investigating this as [for examining] what I otherwise enclosed concerning the problem of longitude.[3] From the *Gentleman's Magazine* for August 1754, in which the announcement of my previous tables was published, I see how much I am indebted to your very kind recommendation concerning the good impression of these tables there [i.e. in England], and return my particular thanks. In the paper referred to, it has been alleged that one has found from comparison with the observations of Halley, that my old tables once erred by about 4′ 37″.[4] In the following issue, however, for September this was contradicted and it was stated that after one had consulted Halley's manuscripts, it appeared that the observation may itself be wrongly reduced and agree correctly with the tables after the necessary improvement [was made].[5] I hope that M. d'Alembert's objection against these very tables will itself be refuted in a similar way. Meanwhile, as to the improved version of these tables,[6] I have preserved them without assuming new equations; except that several of the former [equations] have been somewhat amended. I hope however, that the small error of 30″ can still be reduced much further through [the introduction of] several new equations. Since I have determined rather accurately in these researches those inequalities which depend upon the solar parallax, and among these the one which is most considerable (which has for its argument the distance of the ☽ from the ⊙),[7] I am assured that the parallax of the ⊙ is not greater than $7\frac{1}{2}''$.

As soon as I get my manuscript back from England, I shall hand it over to be printed, and the bookseller Nourse in London has already offered to do this.[8] I shall send to the latter, among other things, the method which I have used for deriving the inequalities of the Moon's motion from theory.[9] This method is quite simple, and correct too; it is in excellent agreement with that according to which I determined the inequalities of latitude and which I have sketched for you in one of my previous communications.[10] Since my formulae are just as well suited for Jupiter as for the Moon, I have also been able to find easily from them the motion of the former with its inequalities;[11] and this theory of Jupiter varies from the observations by not more than about a minute. Now there is still a lack of observations upon which one can sufficiently depend. Yet one would have been able to use the very same [method] to greater advantage by a further improvement of the lunar theory, and since I consider the occultations of the fixed stars by the Moon as the best possible for this purpose, I therefore wish that these might be industriously observed also at the Berlin observatory in particular. The many occultations of Aldebaran which Kirch (more than others) has previously observed there, have been very useful to me; and since the times are now again approaching in which this star can often be obscured by the Moon, I harbour the hope that the astronomers will be particularly attentive to them. Mr Aepinus in Rostock,[12] from whom I learn that he will soon be going to Berlin and will get the use of the observatory there, possesses all that insight and skill which makes him worthy of occupying this position; and I expect a great deal from him, particularly concerning the proposed observations [that are] to be made.

I have the honour to remain with the greatest respect,

Your obedient servant,

T. Mayer

31. *Euler to Mayer; Berlin, 27 May 1755*

I have cause to congratulate you heartily on the fact that your lunar tables found such great acclaim in England, since this nation is very late in giving foreign discoveries their proper right. Only, because all tables constructed upon the Newtonian theory have for so many years erred by over 5′ from the truth, the English self-complacency could not stretch to such a point that the superiority of your tables could remain unknown. What I did about it perhaps lies in the fact that these were investigated industriously and calculations based upon them which, after all, probably did not occur in France where I recommended them with equal emphasis. Since you now have found means of improving them still more, and are in a position to be able to determine through them the longitude at sea, so much is certain; that until now no one has been able to make such a well-established claim on the fixed prize, and since in addition you have solved this problem in the most accurate manner, I heartily wish that this will be recognised in England in accordance with its worth and that the maximum prize will be offered to you; a just decision which would certainly please me very greatly. M. d'Alembert will remain in disgrace in this matter as in many others.

According to this fortunate completed work you seem indeed to have connected your researches made on this subject in a special treatise to lay before the eyes of the world, whereby it would be a credit to the German nation if you would set down beforehand a complete history of the previous efforts of determining the Moon's motion. That the solar parallax does not exceed $7\frac{1}{2}''$, I learn with particular pleasure, since other phenomena had led me also to the conclusion that it must be smaller than $10''$. You will undoubtedly also have worked on the Paris question of next year, since according to these same researches the determination of the inequalities in the motion of the Earth must be smaller.[1] I have discovered from it important phenomena in astronomy; according to which firstly the obliquity of the ecliptic decreases by about $48''$ per century, and secondly the latitude of the fixed stars must have changed, whereby I found an excellent agreement from the old observations. There also arises out of this a new inequality in the precession of the equinoxes and in the longitude of the fixed stars whose difference does not always remain the same as has hitherto been believed. According to this it is quite certain that the obliquity of the ecliptic has previously been larger than now, yet it could have been nowhere near $30° 51'$ at the time of Hipparchus. You will undoubtedly also have discovered the reason why the ancient observations have produced too large a value for the obliquity.

Our Mr Aepinus[2] is taking great pains to put our astronomical instruments into such a state that he can plan really accurate observations with them, for which he does not lack the necessary scientific knowledge and skill. Thus he will soon be in a position to make the lunar observations which you desire, as he is paying particular attention to this.

I congratulate Mr Lowitz heartily on his promotion and beg you to give him my most respectful compliments, as also to Councillor Frantz, if he has already established his home in Göttingen.[3]

In continuation of your highly-treasured friendship I present my special compliments, and have the honour to remain with the greatest respect,

Your obedient servant,

L. Euler

NOTES

Letter 1

1. Georg Moritz Lowitz (1722–74) and Euler appear to have begun corresponding in 1745 and to have continued to do so until the former took up a professorship in astronomy at the St Petersburg Academy in 1767. Nine letters from Lowitz to Euler preserved in the archives of the U.S.S.R. Academy of Sciences refer to Lowitz's work for the Homann Cartographic Bureau in Nuremberg on the preparation of globes and maps, the perfecting of astronomical instruments, the theory of comets and his invitation to St Petersburg on Euler's recommendation.
2. Euler's "Nova theoria lucis et colorum" was written in 1744 and published in his *Opuscula* (*varii argumenti*), i (1746), 169–214.
3. Paris shoe (or foot) = 0·32484 metre.
4. le ≡ $\log_e e$, the natural logarithm of the exponential function.
5. This is probably an allusion to the experiments carried out by the French expedition which was sent to Peru in 1735 for the prime purpose of measuring a length of a degree of the meridian at the equator, the results of which are described in P. Bouguer, *La Figure de la Terre* (Paris, 1749).
6. See Mayer's pamphlet *De refractionibus objectorum terrestrium* (Göttingen, 1751).
7. This implies that 1 Paris shoe ≡ 1·035 Rhein shoes (or 1 Rhein shoe = 0·31385 metre), which differs slightly from the value of 1·030 Rhein shoes adopted by Euler in his *Neue Grundsätze der Artillerie* (Berlin, 1745). See Scherrer's edition (note 12, below), p. 102.
8. Mayer is probably referring to the tracts entitled "Demonstratio Gemina Theorematis Newtoniani . . ." and "Animadversiones in Rectificationem Ellipsis" which appear in the second volume of Euler's *Opuscula* (Berlin, 1750). Cf. pp. 108–20 and 121–66 respectively.
9. "Recherches sur les irregularités du mouvement de Jupiter et de Saturne", *Recueil des pièces qui ont remporté les prix de l'Académie Royal des Sciences, depuis leur fondation*, vii (1769), 1–84.
10. This was the viewpoint of Alexis Claude Clairaut (1713–65) when he wrote his memoir "Du Système du Monde dans les principes de la gravitation universelle", *Histoire de l'académie royale des sciences de Paris. Année MDCCXLV* (1749), *Mémoires*, 329–90; but by 1751 he had come

to accept the inverse-square law of gravitational attraction as a sufficient basis for accounting for the observed motion of the Moon.

11. Mayer may have obtained this information from E. Halley, *Tabulae astronomicae; accedunt de usu tabularum praecepta* (London, 1749). Cf. section entitled "De lunae computo corrigendo" where Halley takes the Saros cycle of 223 Synodic months $\equiv$ 18^y 11^d 7^h 43^m 20^s if there were four leap years, or one day less if there were five, during that interval.

12. B. Robins' *New Principles of Gunnery* (London, 1742) was greatly admired by Euler, who translated it into German and wrote commentaries on most of the sections which it contained. Euler's edition was published under the title *Neue Grundsätze der Artillerie* (Berlin, 1745) and it is this to which Mayer is here referring. Robins' own work was reprinted posthumously by James Wilson in the first volume of the *Mathematical Tracts of the Late Benjamin Robins* (London, 1761), and a corrected and enlarged version prepared by Charles Hutton was published in London in 1805. Euler's treatise, edited by Fredrich Robert Scherrer, was itself reprinted together with four tracts on ballistics in the second series of his *Opera Omnia*, vol. xiv (Leipzig and Berlin, 1822).

13. A considerable quantity of unpublished source material on the siting, construction and instrumentation of this observatory is contained under "Betr. Sternwarte zu Göttingen" (Ph 1/18 4/V f 1–4) and "Personalakte Tobias Mayer" (4/Vb 18) in the University Archive, Göttingen; also, under "Tobias Mayer's Nachlass . . ." (*Cod. MS. Tob. Mayer* 15_4) and "Briefe von Ch. L. Scheidt an Joh. David Michaelis" (*Cod. MS. Philos.* 157) in the University Library, Göttingen. Professor Johann Andreas von Segner (1704–77) was Professor of Mathematics in Göttingen before being called to Halle in 1754, whereupon Mayer assumed sole directorship of the Göttingen Observatory. Little has, however, so far been written about this pre-Gaussian period of astronomy in Göttingen, although a popular article by Horst Michling, "Von der 'alten' Göttingen Sternwarte und ihren Astronomen", has appeared in the *Göttinger Tageblatt* (Nummer 209; Donnerstag, der 8 September 1966).

14. Nicolas Louis de la Caille (1713–62) was Professor of Mathematics in Mazarin College where he had a small observatory. His proposal to undertake a personal voyage to the Cape of Good Hope for the purpose of making astronomical observations of the southern skies was officially sanctioned by the Paris Academy of Sciences, of which he was a member, in 1750 and he duly arrived at that destination in 1753. Besides determining the lunar and solar parallaxes with the aid of the planet Mars, he made the first measurement of an arc of the meridian in the

southern hemisphere and observed some 10,000 southern stars before returning to Paris in 1754.

15. Mayer's results are published in his article "In parallaxin lunae euisdemque a terra distantiam inquisitio", *Commentarii societatis regiae scientiarum Gottingensis*, i (1752), 159–82, which shows that he is assuming the Earth to be spheroidal in form with an ellipticity of 1/220.

Letter 2

1. Euler may have been referring here to the five treatises by Mayer published in the *Kosmographische Nachrichten und Sammlungen auf d.J. 1748* (Nuremberg, 1750). These contain the results of the numerous observations of the Moon's occultations of stars, topographical features, parallax, libratory and sidereal motions, etc., which the latter had been making at Nuremberg in the years 1747 and 1748. He might also have had in mind Mayer's calculations for a lunar eclipse on 23 November 1748 (which he knew through his correspondence with Johann Michael Franz).
2. Cf. pp. 67–105 of the periodical cited in the text, which contains contributions submitted in the years 1747/48 published in 1750.
3. "Conjectura physica de propagatione soni ac luminis", *Opuscula* (*varii argumenti*), ii (1750), 1–22. Cf. ref. 51.
4. This is the length of the Saros cycle of eclipses; that is, of 223 lunations or 19 revolutions of the Sun with respect to a lunar node.
5. The absolutely divergent series $1 - U + 2!U^2 - 3!U^3 + \cdots$ is the asymptotic expansion of the function

$$F(U) = \frac{e^{1/u}}{U}\int_{1/u}^{\infty} \frac{e^{-1/t}}{t}\,dt \quad \text{when } U = 0$$

Putting

$$U = -\frac{1}{\log_e x}$$

yields

$$F\left(-\frac{1}{\log_e x}\right) = \frac{\log_e x}{x}\int_0^x \frac{dt}{\log_e t}$$

The relationship between this divergent series and the continuous fraction given here by Euler is discussed in O. Perron, *Die Lehre von den Kettenbrüchen* (Leipzig, 1929), Ch. 7, §§54, 56 and Ch. 8, §§59, 63.

Letter 3

1. *Historisch und geographischer Calender . . . unter Genehmhaltung der von Sr. Königl. Maj. in Preussen in Dero Residenz Berlin gestifteten Academie der*

Wissenschaften 1750–54 (Berlin, 1749). Mayer is probably referring to the notes in this publication on various physical phenomena in the terrestrial atmosphere.

2. Cf. note 3 to Letter 2.
3. L. Euler, "Dissertatio de magnete ab Illustr. Academia Regia Paris. Scient. praemio condectorata A. 1744", *Opuscula varii argumenti*, iii (1751), 1–53.
4. Gerhardt Müller (1718–62), a garrison doctor in Weimar who subsequently became Professor of Surgery and Botany in Giessen, wrote a dissertation entitled "Untersuchung der wahren Ursache von Newtons allgemeinen Schwere; wie auch bewegenden Kräfte der Körper" (Weimar, 1743). Cf. ref. 50.
5. A. C. Clairaut, *Théorie de la lune déduite d'une seule principe de l'attraction réciproquement proportionelle aux quarrés des distances* (Berlin, 1752). See note 6 to Letter 22.
6. C. Mylius, "Gedenken über des Herrn Prof. Mayers in Göttingen Beweis daß der Mond keinen Luftkreis habe", *Physikalische Belustigungen*, i (1751), 288–302.
7. L. Euler, "Sur l'atmosphère de la lune prouvée par la dernière Éclipse annulaire du soleil", *Histoire de l'académie royale des sciences et belles lettres de Berlin, Année MDMCCXLVIII*, iv, (1750), 103–21.

Letter 4

1. Euler was probably referring to P. Bouguer, "Sur les réfractions astronomiques", *Histoire de l'académie royale des sciences, Année MDCCLI* (1752), Mémoires, 117–57. See, however, note 1 to Letter 5.
2. See note 4 to Letter 3.
3. Euler's *Theoria motuum lunae* (Berlin, 1753) was published by the Berlin Academy soon after Clairaut's *Théorie de la lune . . .* (Berlin, 1752).

Letter 5

1. Mayer was referring to P. Bouguer, "Sur les réfractions astronomiques dans la zone torride", *Histoire de l'académie royale des sciences, Année MDCCXXXIX* (1741), *Mémoires*, 407–23. This may not have been the treatise which Euler mentioned in his previous letter.
2. D. Bernoulli, *Hydrodynamica; sive, de viribus et motibus fluidorum commentarii* (Strassburg, 1738), p. 221.
3. See note 3 to Letter 2.
4. "In parallaxin lunae euisdemque a terra distantiam inquisitio", *Commentarii societatis regiae scientiarum Gottingensis*, ii (1753), 159–82.

Mayer submitted this treatise to that society on 8 April 1752 after having used his manuscript as the basis of a lecture on the subject.

5. D. Bernoulli, "Traité sur le flux et reflux de la mer", *Recueil des pièces qui ont remporté les prix de l'Académie Royale des Sciences*, iv (1752), 55–143.
6. F. Cajori (ed.), *Sir Isaac Newton's Mathematical Principles of Natural Philosophy and his System of the World.* (Berkeley and Los Angeles, 1966); Vol. II, Proposition XXIX. Problem X, pp. 448–9.
7. *MDCCXV*, See note 3 to Letter 4.
8. M. de l'Isle le Cadet, "Réflexions sur ... un anneau lumineux semblable à celui qu'on aperçoit autour de la Lune dans les Eclipses totales du Soleil", *Histoire de l'académie royale des sciences, Année MDCCXV* (Amsterdam, 1719); Mémoires, 220–5.
9. Christlob Mylius had attacked "Tobias Mayers Beweis daß der Mond keinen Luftkreis habe", *Kosmographische Nachrichten und Sammlungen auf d. J. 1748* (Nuremberg, 1750), pp. 397–419. See note 6 to Letter 3.
10. A lengthy theoretical proof of Mayer's formula from Simpson's rule for refraction is given by Christian Bruhns, *Die astronomische Strahlenbrechung in ihrer historischen Entwicklung* (Leipzig, 1861), pp. 51–60. The coefficients, which are derived empirically, do not correspond exactly to those quoted here, but to those in the formula cited in Mayer's *Tabulae motuum solis et lunae novae et correctae* (London, 1770), p. 64.
11. Pierre Charles le Monnier (1715–99) published the first volume of his four-volume work *Observations de la lune, du soleil et des étoiles fixes* in Paris during the previous year (1751). Mayer may have extracted observational data from this, or from either of le Monnier's earlier works, *Histoire céleste* (Paris, 1741) and *Institutions astronomiques* (Paris, 1746).
12. See note 1 to Letter 5 (*op. cit.*).

Letter 6

1. This is an anticipation, by some thirty-five years, of the law which is generally attributed to the French mathematician and physicist Jacques Alexandre César Charles (1746–1823) and called by his name.
2. See note 3 to Letter 4.
3. See note 7 to Letter 3 (*op. cit.*, pp. 118–19).

Letter 7

1. The full reference to this work by Euler is "De machinarum tam simplicium quam compositarum usu maxime lucrose", *Commentarii academiae scientiarum imperialis Petropolitanae*, x (1747), 67–94.

2. Some of these were embodied by Mayer in a course of public lectures on mechanics which he began on the very same day as he wrote this letter. He discusses the subject of water-wheels in lectures delivered between 1 and 5 September 1752. For details, see *Cod. MS. T. Mayer* 15_{34} (IV Theil, 2. Cap. §§266–71), in the Göttingen University library.

Letter 8

1. This statement is a clear indication that one of Euler's letters is still missing.
2. Euler had probably referred Mayer to his article "Recherches sur l'effet d'une Machine Hydraulique proposée par M. Segner, Professeur a Gottingue", which was published in the *Histoire de l'académie royale des sciences et belles lettres de Berlin Année MDCCL* (1752), 311–54. The "further explanations" to which Mayer refers in his opening sentence could refer both to ideas contained in this and in another article by Euler on the same theme: "Application de la Machine hydraulique de M. Segner, à toutes sortes d'Ouvrages, et de ses avantages sur les autres machines hydrauliques, dont on se sert ordinairement"; for the latter, which appears on pp. 271–304 of the same periodical for the year 1751 (published in Berlin in 1753), would have been written by this time.

Letter 9

1. J. Cassini, "Des réfractions astronomiques", *Histoire de l'académie royale des sciences de Paris, Année MDCCXIV* (1717); Mémoires, 33–54.
2. L. Euler, *Theoria motuum lunae . . .* (Berlin, 1753).
3. A. Clairaut, *Théorie de la lune déduite d'un seul principe de l'attraction* (Berlin, 1752).
4. This is probably a reference to the volume for the year 1750, published in 1752, which contains three treatises by Euler and Segner's description of his hydraulic machine.
5. This was wishful thinking on Mayer's part. The building itself was not completed until the autumn of 1754, and a further year was to elapse before its major instrument—a 6-ft radius mural quadrant constructed by the London instrument maker John Bird—was installed. The first reliable observations of the equatorial co-ordinates of fixed stars were made by Mayer, using this instrument in conjunction with a pendulum clock, in August 1756.
6. A description of this instrument is given by Mayer in "Nova methodus perficiendi instrumenta geometrica. Et novum instrumentum gonio-

metricum", *Commentarii societatis regiae scientiarum Gottingensis*, ii (1753), 325–36. A German translation of this article is given by Johann Friedrich Benzenberg in the introduction to his edition of the *Erstlinge von Tobias Mayer* (Düsseldorf, 1812).

7. This is probably a reference to the clocks and 3-ft radius quadrant which F. L. Kampe, a skilled local artisan, had been commissioned by Professor Segner to construct for the Observatory. The problem of obtaining suitable brass and Kampe's responsibilities as a Senator on the Göttingen town-council caused delays which had not been anticipated; thus, although Kampe managed to complete the clocks to Segner's satisfaction sometime before 16 November 1752 he never did complete the quadrant and Mayer finished the work himself. For details, see for example "Betr. Sternwarte zu Göttingen" Ph.1/18 4/V f/1 (items 104, 129–31, 133) and *Cod. MS. Tob. Mayer* 15_4 (items 45, 58, 59, 60, 62) in the University archive and library, Göttingen, respectively.
8. Pieter van Musschenbroek (1692–1761), Professor of Mathematics and Physics at Duisburg, Utrecht and Leiden. The book which Mayer is most likely to have consulted is: *Hrn Peters von Musschenbroek . . . Grundlehren der Naturwissenschaften, nach der zweyten lateinischen Ausgabe* (Leipzig, 1747); edited by J. C. Gottscheden.
9. Georges Buffon (1707–88), Director of the Botanic Garden in Paris, had discussed the subject in question in an article entitled "Expériences sur la forêe du bois", *Histoire de l'académie royale des sciences de Paris, Année MDCCXL* (1742); Mémoires, 453–67.
10. Galileo Galilei, *Discorsi e dimostrazione matematichè intorno à due nuove scienze* (Leiden, 1638). See Henry Crew's English translation (1914), Second Day, pp. 151–89.

Letter 10

1. This remark has been interpreted by Dr Kopelevich to mean that Letter 9 was never sent, implying that Mayer erred by about three months in recalling the date on which he wrote Letter 8. In view of the uncertainty in dating Letter 9, however, one ought not to discount the alternative possibility that the latter—or an improved version of it—was indeed dispatched by Mayer sometime in late July or early August 1752.
2. See note 2, Letter 8. The history of Segner's invention of a water-turbine (the Segner Wheel) and the important part played by Euler in it, are revealed by the correspondence between these two men from 1750 to 1752 preserved in the Archive of the Academy of Sciences, U.S.S.R., Fol. 136, Op. 2, Nr. 3.

3. This is a reference to the tables of John Flamsteed (1646–1719), the first English Astronomer Royal, which were published posthumously in his *Historia coelestis Britannica*, 3 volumes (London, 1725).

Letter 11

1. The President of the Berlin Academy at that time was Pierre de Maupertuis (1698–1759), the French mathematician and astronomer who had led the expedition into Lapland in 1736 to measure the length of a degree of the meridian but better remembered by posterity as the originator of the principle of Least Action which bears his name.
2. In Mayer's letter of appointment, his basic salary was fixed at 400 Reichsthaler (cf. "Personalakte Tobias Mayer" 4/Vb 18, p. 5, preserved in the University archive, Göttingen). He also received "Licent-Aequivalent-Gelder" of 40 Rthl. per annum and augmented his salary further through private teaching (*ibid.*, pp. 2, 7, 11, 12).

Letter 12

1. This phrase may be significant for two reasons: it could refer to Mayer's personal intention to buy a house as he did towards the end of the following year after he had made up his mind to remain in Göttingen; or, to the common intention of Mayer, J. M. Franz, and G. M. Lowitz to transfer the Cosmographical Institute and its Globe-Factory from Nuremberg to Göttingen. This idea had originated more than three years previously in Franz's reply to the suggestion of C. L. Scheidt (a Hanoverian State Councillor) that a similar transfer be made to Berlin (Ph. 1/18 4/V f/1, nr. 15, University Archive, Göttingen).
2. "Novae tabulae motuum solis et lunae", *Commentarii societatis regiae scientiarum Gottingensis*, ii (1753), 383–430.
3. Christopher Schumacher (1704–68) brought Mayer a present from Euler of observations made at Greenwich by the Astronomer Royal James Bradley, which Euler had previously received via Gael Morris. These are the lunar observations to which Mayer is here referring. They are appended to Mayer's "Tabularium lunarium in Commentt. S.R. Tom. II contentarum usus in investiganda longitudine maris", *loc. cit.*, iii (1754), 375–84 (cf. pp. 393–6).
4. A catalogue of lunar eclipses giving details of 64 eclipses between 5 July 1610 and 17 April 1753, arranged in order of their respective Saros cycles, is appended to *ibid.*, pp. 385–92.
5. R. Dunthorne, "On the motion of the moon", *Philosophical Transactions*, xliv (1747), 412–20; and "On the acceleration of the moon", *loc. cit.*, xlvi (1752), 162–72.

6. John Flamsteed's Greenwich observations from 1676–1705 were included in the pirated two-volume *Historia coelestis* (London, 1712). All his observations from 1675–1724 were contained in the posthumously-published three-volume *Historia coelestis Britannica* (London, 1725).
7. E. Halley, *Astronomical Tables* (London, 1752).
8. See note 9 to Letter 1.

Letter 13

1. See note 3 to Letter 12.

Letter 14

1. See note 9 to Letter 1.
2. Johann Kepler first enunciated the law that the cube of the mean distance between a planet and the Sun is proportional to the square of its period of revolution around the Sun, in his *De Harmonice Mundi* (Augsburg, 1619).
3. W. Gardiner, *Tables of logarithms for all numbers, from 1 to 102,100 and for lines and tangents to every ten seconds of each degree in the quadrant* (London, 1742).

Letter 15

1. See note 3 to Letter 14.
2. Fieldmarshall von Schmettau was the instigator of the suggestion to move the Homann Cartographic Bureau from Nuremberg to Berlin, in order that the Berlin Academy could benefit from the expertise of its workers in obtaining proper engravings of maps which would be necessary to effect that academy's own plans to prepare an atlas of Germany (for details, see *Cod. MS. Tob. Mayer*, 15_2, p. 6, in the Göttingen University library). His letter to this effect, written to Gerlach Adolph von Munchhausen at the beginning of 1750, was directly responsible for the latter instructing Ch. L. Scheidt to test Franz's reaction to this proposal (cf. note 1 to Letter 12).

Letter 16

1. L. Euler, *Theoria motuum Lunae* . . . (Berlin, 1753).
2. Details of eclipse observations up to the year 1678 are contained in an unpublished quarto manuscript entitled "Historia eclipsium ab anno 1610, quo telescopio observationes fieri coeperunt". (*Cod. MS. Tob. Mayer*, 15_{48}, in the Göttingen University Library).

3. Mayer wrote a treatise entitled "Methodus facilis et accurata computandi eclipses solares in dato loco conspicuas", which he read to the Göttingen Scientific Society on 3 September 1757. The full Latin text is published in G. C. Lichtenberg, *Opera inedita Tobiae Mayeri I* (Göttingen, 1775), pp. 23–30, although a German abstract appeared in the *Göttingische Anzeigen von gelehrten Sachen* (1757), 1065–6.
4. See ref. 32 (p 25.)
5. The local artist was the Göttingen Senator F. L. Kampe. See note 4 to Letter 26.

Letter 17

1. Mayer is here referring to Euler's idea that the distances of the planets from the Sun are gradually diminishing on account of the resistance of an aetherial fluid; and he may have read the extract from Euler's letter to Caspar Wetstein (28 June 1749) on this matter, published in the *Philosophical Transactions*, xl (1752), 203.
2. Hipparchus was a Greek astronomer born at Nicaea in Bithynia early in the 2nd century B.C., who observed and catalogued more than 1000 stars from some site on the island of Rhodes. He is famous for his discovery in 130 B.C. of the phenomenon of precession of the equinoxes.
3. Claudius Ptolemaeus (Ptolemy) was a Greek mathematician, astronomer and geographer whose *Almagest* contains astronomical observations made at Alexandria over a period of many years ending in A.D. 127. This work, which has been the subject of numerous Arabic and Latin commentaries throughout the medieval period, is an admixture of Aristotelian physical cosmology and the geometrical constructions of Greek planetary theory.
4. Albategnius (858–929), alias Al-Battani, was an Arab astronomer born in Batan in Syria whose principal work *De motu stellarum*, was published at Nuremberg in 1537 by Philipp Melanchthon (1497–1560).
5. Philip van Lansberg (1561–1632), Dutch theologian and astronomer, and a strong adherent of Copernican astronomy. He preferred the latter's geometrical constructions to Kepler's laws of planetary motion, basing astronomical ephemerides on them in 1632. This is likely to have been the aspect of Lansberg's work which Mayer is criticising in this present context.
6. Giovanni Battista Riccioli (1598–1671), whose lunar maps—Figs. 1–6 in his *Almagestum Novum* . . . (Bologna, 1651)—had been found by Mayer to be very unreliable as a basis for the eclipse method of longitude determination.

Letter 18

1. This remark would appear to refer to the content of Euler's reply to

Mayer's interesting letter of 22 August 1753, which unfortunately has not been preserved.

2. P. C. le Monnier, *Histoire céleste, ou recueil de toutes les observations astronomiques faites par ordre du roy* (Paris, 1741).
3. The phenomenon of nutation was discovered by the English Astronomer Royal James Bradley, and publicly announced on 14 February 1747 (*Philosophical Transactions* xlv, 1–43). Jean d'Alembert was the first person to rigorously develop the theory of nutation (and precession) from the principles of Newtonian attraction. His work on this subject, which Mayer had probably read, was published in Paris in 1749.
4. Bernhard Walther (1440–1504) was a student of Regiomontanus in Nuremberg. The observations of both these astronomers were published by Walther's younger collaborator Johannes Schoner in 1544.
5. This difference method was employed by Mayer in his treatise "Latitudo geographica urbis Norimbergae", *Commentarii societatis regiae scientiarum Gottingensis,* i (1752), 373–8.
6. See note 2 to Letter 17.
7. J. Cassini, *Élémens d'astronomie* (Paris, 1740).
8. Jacob Adami, a Prussian amateur mathematician, began corresponding with Euler in 1746; and three years later he sent the Berlin Academy an essay entitled "Specimen hydrodnamicum de resistentia corporum in fluidis motorum". The decision to award Adami a prize for this contribution was announced at a public meeting of the Berlin Academy on 1 June 1752. It was subsequently reviewed in the *Göttingische Anzeigen von gelehrten Sachen* (1753), 935–6.
9. Abraham Gotthelf Kaestner (1719–1800), at that time Professor of Mathematics at Leipzig, who took up a similar appointment at the Georg-August Academy in 1756 and remained there until the time of his death.
10. This is a map of the military roads of the Western Roman Empire.

Letter 19

1. Euler is probably referring to volumes I and II of the Göttingen *Commentarii*, and not to vol. III as Dr Y. K. Kopelevich has suggested in her note 1 to this letter in the 1959 Russian edition of the Euler–Mayer correspondence (cf. Preface); for Mayer refers to vol. III almost four months later as being recently published (cf. Letter 22).
2. Euler himself had obtained this lengthy (composite) letter, which was a copy of the original from the Jesuit missionary Father Gaubil in Peking

to Joseph de l'Isle in Paris, from his correspondent Caspar Wetstein in London (cf. Archive of the Academy of Sciences, U.S.S.R. fol. 136, op. 2, no. 22, pp. 36–8). It is still preserved among other letters to Mayer in *Cod. MS. T. Mayer* 15_2 in the Göttingen University library, where it is catalogued as: "Bl. I. Vier Briefe von A. Gaubil J., Peking 22 Oct. und 9 Nov. 1752, 'écrites . . . à mons. De L'Isle . . . à Paris', Copie." In his unpublished "Vorlesungen über Sternkunde" (*Cod. MS. T. Mayer* 9), written in 1753, Mayer uses his solar and lunar tables to check the reliability of Gaubil's prediction of a particular solar eclipse supposed to have occurred in China in the third millenium B.C., and rightly rejects it as fictitious.

Letter 20

1. *Ephemerides motuum coelestium, ex. anno 1751 in annum 1762 ad meridianum Bononiae supputatae; auctoribus Eustachio Zanotto, Bononiensis scientiarum Instituti astronomo, et sociis, ad usum Instituti. Praemissae sunt novae tabulae motus Jovis* (Bologna, 1750).
2. See note 8 to Letter 12.
3. See note 2 to Letter 19.

Letter 21

1. The letter from Count Razumowsky to Euler has not been preserved, but the question of inviting Mayer to St Petersburg is discussed in correspondence between G. F. Müller and Euler published in Volume 1 of *Die Berliner und die Petersburger Akademie der Wissenschaften im Briefwechsel Leonhard Eulers* (Berlin, 1959).
2. Elizabeth [Petrovna] (1709–62), Empress of Russia, the daughter of Peter the Great and Martha Skovronskaya.
3. This does not refer to a separate individual but is a military title which applies to The Lord President himself. The word "hetman" (or "ataman") was introduced into Russia from Poland and adopted by the Cossacks as a title for their head or elected elder of the "stanitsa", the unit of Cossack administration.
4. See note 2 to letter 12.
5. J. le Rond d'Alembert, *Recherches sur différents points importants du système du monde* (Paris, 1754). In this treatise, whose publication was completed in 1756, d'Alembert perfected the solution of the problem of planetary perturbations.
6. See, for example, E. G. Forbes, "Tobias Mayer's Lunar Tables", *Ann. Sci.*, xxii (1966), 105–16.

Letter 22

1. Albrecht von Haller (1708–77), Swiss-born professor of medicine, anatomy, botany and surgery at the Georg-August Academy in Göttingen from the time of its foundation (1737) until 1753, when he resigned his chair and returned to Berne, his birth-place.
2. B. Stähelin (1695–1750), professor of physics at Basle University.
3. See note 3 to Letter 12.
4. Certainly a reference to A. Clairaut, "De l'orbite de la lune dans le système de Newton", *Histoires de l'académie royale des sciences de Paris, Année MDCCXLIII* (1746); Mémoires, 17–32.
5. Three letters from Mayer to Joseph De l'Isle, dated Nuremberg 21 August and 6 November 1748 and 14 January 1751, are preserved among the latter's unpublished correspondence at the Observatoire de Paris. They are in tome IX no. 178, tome X, no. 45 and tome XI, no. 57 respectively. Despite the discrepancy in the years cited here by Mayer, these are undoubtedly the letters to which he is referring.
6. Yu. Kh. Kopelevich, "The Petersburg Astronomy Contest in 1751", *Soviet Astr.—AJ*, ix (1966), 653–60.
7. Mayer subsequently prepared a catalogue of zodiacal stars which was published in G. C. Lichtenberg, *Opera inedita Tobiae Mayeri* I (Göttingen, 1775), pp. 49–74. However, this does not contain the results of the observations which he made prior to 1756 (to which he is here referring).
8. This is Mayer's "Methodus longitudinum promota", to which he later added an appendix. The method was published posthumously by Nevil Maskelyne together with Mayer's *Tabulae motuum solis et lunae novae et correctae* (London, 1770).

Letter 24

1. Johann Kiess (1713–81), Professor of Mathematics and Physics and Director of the Observatory at Berlin (1742–54), thereafter Professor of Mathematics at the *Collegium illustre* in his native Tübingen, was a firm adherent to Newtonian ideas.
2. Georg Wolfgang Krafft (1707–54) was a close friend of Euler's since he had been professor of theoretical and experimental physics at the St Petersburg Academy between 1731 and 1744 before taking up a similar appointment in Tübingen during that latter year.
3. The original letter from Pierre von Maupertuis to Mayer, dated Berlin 27 August 1754, is preserved among the "Briefe von und an J. Tobias Mayer" (*Cod. MS. philos. 159, p. 7*) in the Göttingen University library. It merely confirms what Euler has written in this letter.

4. Georg Moritz Lowitz (1722–74) did not resolve on this occasion to go to St Petersburg, although he took up a professorship in astronomy there in 1763 after having spent the previous eight years as a professor in Göttingen. (See note 3 to Letter 31.)

Letter 25

1. See note 1 to Letter 24.
2. Gottfried Heinsius (1709–69), formerly professor of astronomy at the St. Petersburg Academy, moved to Leipzig in 1744. There is evidence in the Protocols of the St Petersburg Academy Conference for 18 August 1754 that he was also assisting that academy to fill its vacant posts.

Letter 26

1. Mayer's letter of 2 September 1754 to von Munchhausen is contained among his *Personalakte* (4/Vb 18 no. 16) in the University archive, Göttingen.
2. Christfried Kirch (1694–1740), who had taken over his father's position as professor of astronomy at Berlin in 1717, died leaving a large amount of manuscript material of his observations in the Berlin Observatory.
3. The following sentence makes it clear that Mayer is referring to his manuscript tables incorporating the improvements which he had described in his letter of 6 March 1754 (Letter 20), and not the improved tables which had been published in Volume III of the Göttingen *Commentarii*.
4. This was a wooden model of Mayer's Repeating Circle, which F. L. Kampe promised to make. His onerous duties on the Göttingen town-council were largely responsible for his failure to complete this instrument by the time Mayer wanted to send it to the Admiralty in London (three months later) in support of his claim for the longitude prize offered under the terms of the Act 12 Anne (1714); with the result that Mayer graduated it himself and duly transmitted it to London via the normal diplomatic channels in December 1755.
5. The above remarks constitute a summary by Mayer of the principles contained in his "Methodus longitudinum promota" which, in its original form, was in the hands of the English Astronomer Royal James Bradley by 9 November 1754 (see note 8 to Letter 22).
6. The story of the difficult situation which was soon destined to arise as a result of the transfer of Lowitz's globe factory from Nuremberg to Göttingen, and his failure to complete the task of making these globes,

is described by Sophus Ruge, "Aus der Sturm-und-Drang-Periode der Geographie", *Zeitschrift für Wissenschaftliche Geographie*, v (1885), 249–60 and 355–64 (see pp. 357–62).

Letter 27

1. A rough draft of Mayer's reply to the offer from the St Petersburg Academy, addressed to G. F. Müller, is preserved on the same page (p. 50) of *Cod. MS. philos. 159* in the Göttingen University library as that containing the draft of this letter to Euler.

Letter 28

1. Mayer's relationship with his co-director Professor Segner at this time was not a friendly one, owing to some misunderstanding during the construction of the Göttingen observatory which had given rise to feelings of mutual distrust between them. Had this not been the case, Mayer would never have asked to be given sole charge of the observatory (cf. *Cod MS. philos. 159*, p. 65, Göttingen University library) which, for von Munchhausen, was the most embarrassing of all the conditions made by Mayer (cf. letter from Ch. L. Scheidt to J. D. Michaelis, 23 September 1754, *Cod. MS philos. 157* in the Göttingen University library).

Letter 29

1. This is certainly an allusion to Segner, whom Mayer knew to be in correspondence with Euler at this time.
2. Although Euler's letter of 22 October 1754 has unfortunately not been preserved, a letter from G. F. Müller, St Petersburg, written in German and dated 17 December 1754 reveals that the latter wished to keep in contact with Mayer and obtain information from him on cartographical matters. (*Cod. MS. philos. 159*, pp. 30, 31 in the Göttingen University library).
3. The meaning of this sentence is apparent from the following remarks by Müller, translated from his letter of 17 December 1754 (see note 2, above): "I am taking the liberty of asking you to inform me about the achievements and present state of the Cosmographical Society. Prof. Euler has probably told you of my desire to obtain a collection of all old and new maps of the Russian State, as many as I can obtain, and also the maps of Poland and Lithuania published by yourself. I would like to repeat this request yet again. . . .
4. These are probably the three maps of the Ukraine published in 1648, 1650 and 1660 by the French engineer G. B. de Vasser.

5. "Essai d'un explication physique des couleurs engendrées sur des surfaces extrêmement minces", *Histoire de l'académie royale des sciences et belles lettres, Berlin. Année MDCCLII*, viii (1754), 262–82.

Letter 30

1. James Bradley (1693–1762), Astronomer Royal, whose opinion was probably known to Euler through his correspondence with Wetstein.
2. Mayer's memorial to the Lords of the Admiralty, his new lunar tables (retrieved from the printer of the Göttingen *Commentarii*), and a manuscript copy of his "Methodus longitudinum promota" were dispatched from Göttingen by Michaelis on 8 December 1754 and were received by W. P. Best in London less than two weeks later. Mainly on account of the Christmas holidays, a further month was to elapse before these papers were finally put into Lord Anson's hands.
3. Namely, the "Methodus longitudinum promota" (see note 5 to Letter 26).
4. Mayer's tables, and Halley's observations, are cited on pp. 374 and 376 respectively of the publication in question.
5. *Gentleman's Magazine*, September 1754, p. 439.
6. That is, the version which was sent to the Admiralty in December 1754 (see note 2, above).
7. This is Mayer's equation XIII (see text of Letter 10).
8. John Nourse wrote to Mayer (in French) on 29 October 1754 requesting his permission to reprint his lunar tables incorporating certain corrections which Euler had pointed out in a letter to Wetstein, with additional rules for deducing the equatorial co-ordinates from the observed celestial ones and for finding the difference in terrestrial longitude from an observed lunar distance. (*Cod. MS. philos. 159*, p. 34 in the Göttingen University library.)
9. This is Mayer's *Theoria lunae juxta systema Newtonianum* (London, 1767), which was edited by Nevil Maskelyne on the instructions of the commissioners of longitude, printed by W. Richardson and S. Clark, and published by Messrs Nourse, Mount and Page.
10. See Mayer's letter of 6 March 1754 (Letter 20).
11. There are about 60 pages of calculations in a folio manuscript entitled "Theoria Jovis" preserved among Mayer's unpublished papers in the Göttingen University library (*Cod. MS. Tob. Mayer*, 15_{29}).
12. Franz U. T. Aepinus (1724–1802), at this time a lecturer in the University of Rostock, was appointed professor of astronomy at the Berlin Academy in 1755 after it was known that Mayer would not be taking up this post. Only two years later, however, Aepinus left Berlin when offered a professorship in physics in the St Petersburg Academy.

Letter 31

1. The subject for 1756 proposed by the Paris Academy was the theory of the inequalities in the motion of the Earth produced by the other planets. The prize was won by Euler himself for his treatise "De motu corporum coelestium a viribus quibuscunque perturbato", *Novi commentarii academiae scientiarum Petropolitanae*, iv (1752/3), 161–96 (published in 1758).
2. See note 12 to Letter 30.
3. Lowitz had been appointed professor of "Cosmography and practical mathematical sciences" in Göttingen and had taken up this post in February 1755. Franz's appointment as Ordinary Professor of Geography at Göttingen, however, was not officially authorised until 21 June 1755 and at the time of Euler's writing this letter he had still not left Nuremberg.

INDEX